Peggy Freudenberg, Sabine Hoffmann (Eds.)
Assessing the Overheating Risk of Buildings

Also of interest

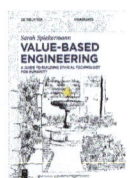

Value-Based Engineering
A Guide to Building Ethical Technology for Humanity
Sarah Spiekermann, 2023
ISBN 978-3-11-079336-9, e-ISBN 978-3-11-079338-3

Thermal Analysis and Calorimetry
Versatile Techniques
Edited by Aline Auroux, Ljiljana Damjanović-Vasilić, 2023
ISBN 978-3-11-059043-2, e-ISBN 978-3-11-059044-9

Thermal Analysis and Thermodynamics
In Materials Science
Detlef Klimm, 2022
ISBN 978-3-11-074377-7, e-ISBN 978-3-11-074378-4

Non-equilibrium Thermodynamics and Physical Kinetics
Halid Bikkin, Igor I. Lyapilin, 2021
ISBN 978-3-11-072706-7, e-ISBN 978-3-11-072719-7

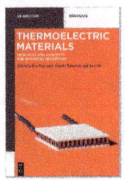

Thermoelectric Materials
Principles and Concepts for Enhanced Properties
Edited by Ken Kurosaki, Yoshiki Takagiwa, Xun Shi, 2021
ISBN 978-3-11-059648-9, e-ISBN 978-3-11-059652-6

Assessing the Overheating Risk of Buildings

Edited by
Peggy Freudenberg, Sabine Hoffmann

DE GRUYTER

Editors
Dr.-Ing. Peggy Freudenberg
TU Dresden
Faculty of Architecture
Zellescher Weg 17
01069 Dresden
Germany
peggy.freudenberg@tu-dresden.de

Prof. Dr. Sabine Hoffmann
RPTU Kaiserslautern-Landau
Faculty of Civil Engineering
Paul-Ehrlich-Str. 14
67663 Kaiserslautern
Germany
Sabine.hoffmann@rptu.de

ISBN 978-3-11-131802-8
e-ISBN (PDF) 978-3-11-131865-3
e-ISBN (EPUB) 978-3-11-131968-1
DOI https://doi.org/10.1515/9783111318653

We acknowledge support for the open access publication by the Open Access Publication Fund of
Saxon State and University Library Dresden (SLUB).

Library of Congress Control Number: 2024941870

Bibliographic information published by the Deutsche Nationalbibliothek
The Deutsche Nationalbibliothek lists this publication in the Deutsche Nationalbibliografie; detailed
bibliographic data are available on the Internet at http://dnb.dnb.de.

Foreword

The issue of overheating in buildings is a critically neglected aspect of building physics. Unlike energy-efficient construction, the consequences of disregarding this planning aspect cannot be directly monetarily quantified, at least in naturally ventilated resp. free-running buildings. Instead, it manifests as a 'silent' economic damage, evident in significant productivity losses among employees in overheated buildings and posing considerable health risks to vulnerable groups, culminating in mortality and morbidity increase. This silent economic impact is particularly pronounced in Europe, where the majority of buildings are not equipped with air conditioning systems, making the mitigation of overheating largely dependent on passive measures.

Amidst the backdrop of climate change, the relevance of this topic intensifies, particularly in Europe, where a notable exacerbation of summer conditions is already observed. Urbanization, coupled with a historical oversight of urban climatic measures, sharply intensifies the phenomena. As cities continue to expand, with an expected urbanization rate of 80% by 2050 in Europe[1], the proliferation of concrete and asphalt contributes to the urban heat island effect, turning urban environments into veritable heat traps during the summer months. This choice of building materials, combined with unfavourable architectural trends such as limited green spaces and large glass façades, further deteriorates the resilience of the urban climate. Cities like London have lost about 10% of their green spaces since 1990[2]. These trends are not sustainable and negatively impact the robustness of urban environments. Additionally, anthropogenic heat sources within cities, such as traffic and the increasing use of air conditioning systems, further exacerbate these unwanted urban conditions. The buildings are not prepared for these challenges. Overheating is a rising concern that is causing more and more deaths. Analyses of the year 2022 revealed more than 60,000[3] heat-related deaths in Europe alone. Our building culture should respond to this development and dedicate more attention to climate stability and protection against overheating. Essential questions should be answered in scientific dialogue, for example the question of tolerance limits for different groups of people, as there are many approaches for assessing comfort, but no minimum requirements to ensure that buildings are at least bearable in summer and do not pose a health risk,

---

1 United Nations, Department of Economic and Social Affairs, Population Division (2018). *World Urbanization Prospects: The 2018 Revision*, Online Edition, https://population.un.org/wup, 2018.

2 E. Ivits, G. Prokop, G. Tóth, M. Gregor, R. Milego, J. Fons Esteve, A. I. Marín, Ch. Schröder, E. Mancosu. *Land take and land degradation in functional urban areas*. European Environment Agency (EEA)- Report No. 17. 78 pages, doi:10.2800/714139. 2021.

3 Ballester, J., Quijal-Zamorano, M., Méndez Turrubiates, R.F. et al. *Heat-related mortality in Europe during the summer of 2022*. Nat Med 29, 1857–1866 (2023). https://doi.org/10.1038/s41591-023-02419-z

especially for sensitive occupant groups. Furthermore, in view of the large number of comfort assessment approaches, the question also arises as to which target corridors should be set for the indoor climate in summer and in what form special occupant groups should be considered in these limit values. The necessity for a stronger orientation of buildings towards the needs of occupants was clearly explained in the recent publication 'Occupant-Centric Simulation-Aided Building Design' by F. Tahmasebi and W. O'Brian[4].

In addition, many questions arise for the summer case assessment of buildings regarding the verification itself, for example, which climatic boundary conditions should be selected to represent adequate conditions of the local climate, e.g., in the form of the Urban Heat Island effect, or the future climate. The selection should not be focused solely on temperature conditions. Meteorological evaluations for Europe show that the summer phases are subject to trends in several weather elements, for example significantly increased solar radiation[5].

Further questions arise in the modelling of summer scenarios for simulation software, as it is often unclear which software includes which model scope or when a detailed representation of individual influencing factors, such as realistic natural ventilation, is necessary. Even in the evaluation of simulation results, much remains unclear, such as the question of which parameters, tolerance limits, and periods should be considered.

This book, *Assessing the Overheating Risk of Buildings*, focuses on the overheating protection of buildings in Europe. It is designed to support the simulation-based evaluation and design of buildings. It addresses the climatic conditions, their changes for Europe and European cities, and the influence of these conditions on the overheating risk of buildings (Chapter 1). It proposes an approach for assessing indoor climate tolerability for European conditions and evaluates known comfort models for their applicability under European summer conditions (Chapter 2). It examines simulation models with a focus on European solar radiation conditions, window and shading systems, the representation of larger thermal masses in historic buildings, and the depiction of natural ventilation (Chapter 3). Finally, it addresses the evaluation of simulation results, focusing on the applied threshold values, the considered time windows, and the permitted tolerances, with reference to existing European standards for assessing overheating risks (Chapter 4).

---

4 O'Brian, W., Tahmasebi, F. *Occupant-Centric Simulation-Aided Building Design: Theory, Application, and Case Studies*. First edition. United States: CRC Press, Taylor & Francis, 399 pages, https://library.oapen.org/handle/20.500.12657/62515, 2023.
5 Yuan, M., T. Leirvik, and M. Wild, 2021: *Global Trends in Downward Surface Solar Radiation from Spatial Interpolated Ground Observations during 1961–2019*. Journal of Climate, 34 pages, 9501–9521, https://doi.org/10.1175/JCLI-D-21-0165.1.

This book represents a pioneering step toward systematically capturing the state of knowledge in this extensive field. It serves to close a gap in the scientific literature and to establish a community on this topic. Consequently, this first version is not intended to be comprehensive. As an Open-Access publication, it is designed to be readily available to a broad professional community, fostering the hope that it will incorporate additional perspectives from active researchers in future revisions. Topics such as mitigation strategies through building design and conservative technical solutions are potential areas for continuation of this project. Other approaches include the diverse modelling approaches in simulation tools, for which there is currently a lack of overarching quality criteria.

I would particularly like to thank the researchers who have already supported this book project with a diligent chapter review and valuable suggestions. These are Marcel Schweiker, RWTH Aachen University, and Farhang Tahmasebi, University College London (Chapter 2), Wilhelm Kuttler, University of Duisburg-Essen (Chapter 1) and Heiko Fechner, Dresden University of Technology (Chapter 3).

I am especially grateful to the authors Christoph Schünemann, Sabine Hoffmann, Abolfazl Ganji and Tim F. Kriesten for their significant contributions.

I would also like to invite the rest of the community to engage with and expand upon the research and discussions presented herein, fostering a collaborative approach to tackling the challenges of building overheating.

Peggy Freudenberg, June 2024

Contents

Peggy Freudenberg, Christoph Schünemann, Tim Felix Kriesten
Climate Conditions and Overheating Risk —— 2

Peggy Freudenberg
Thermal Bearability and Thermal Comfort —— 34

Sabine Hoffmann, Abolfazl Ganji, Peggy Freudenberg, Christoph Schünemann
Building Simulation for Overheating Risk Evaluation and Optimization —— 58

Christoph Schünemann, Peggy Freudenberg, Tim Felix Kriesten
Indoor Overheating Assessment —— 82

Peggy Freudenberg, Christoph Schünemann, Tim Felix Kriesten

Climate Conditions and Overheating Risk

Evaluation of Climate Data's Role in Building Overheating Assessments in Europe

Abstract: This chapter explores the dynamic relationship between climate conditions and the risk of indoor overheating across Europe, with a specific focus on the distribution of hot days and tropical nights. It delves into how geographical factors such as latitude, proximity to the sea, and altitude influence local temperature and radiation conditions, shaping distinct microclimates that affect thermal comfort within buildings. Urban climatic features are discussed to underscore the complexities these environments present, particularly in terms of modifying local climates which exacerbate heat risks. The chapter further addresses the challenges in selecting appropriate climate datasets for overheating risk assessment. It illustrates these challenges through an analysis of weather dataset characteristics, highlighting the ambiguity in determining which dataset might be most critical for risk evaluation. The impact of using different climate datasets on the perceived overheating risk of a specific example building is also examined. Various reference datasets for building simulation were compared, along with different approaches to modelling urban climatic effects and implementing future climate scenarios. This analysis reveals significant variations in risk assessment outcomes depending on the chosen dataset, thereby emphasizing the need for careful selection and consideration of local climatic nuances in building design and urban planning to mitigate overheating risks effectively. This comprehensive approach not only helps in understanding the broader implications of climate change but also guides stakeholders in making informed decisions to enhance indoor environmental quality and occupant health.

Keywords: Climate Change, Climate Conditions, Heat Waves, Urban Heat Island Effect (UHI), Indoor Overheating Risk

Peggy Freudenberg, Dresden University of Technology, Institute of Building Climatology, Zellescher Weg 17, 01062 Dresden, +49(0)351 463-35259, peggy.freudenberg@tu-dresden.de
Christoph Schünemann, Leibnitz Institute of Ecological Urban and Regional Development (IOER), Weberplatz 1, 01217 Dresden, +49(0)351 4679-194, c.schuenemann@ioer.de
Tim Felix Kriesten, Leibnitz Institute of Ecological Urban and Regional Development (IOER), Weberplatz 1, 01217 Dresden, +49(0)351 4679-194, t.kriesten@ioer.de

1 Introduction

The increasing urgency of discomfort, loss of productivity, and health risks due to overheating in European buildings has escalated into a significant concern. Amidst the backdrop of climate change, the intensification and prolongation of heatwaves underscore the necessity for both existing and new buildings to be heat resilient. This resilience, or more precisely, the assessment of overheating risks, is often determined through building performance simulation (BPS), which includes detailed modelling of building physics based on climatic boundary conditions. The *choice of meteorological data for BPS* poses a unique challenge. The data must represent a long-term typical summer for the building's location, yet verified measurements at the building site are typically unavailable, and meteorological conditions, especially wind, can vary significantly. Common practices involve using data from meteorological stations typically located outside urban areas or standardized synthetic data such as Test Reference Years (TRY) for Germany, Design Summer Year (DSY) for the UK, or IWEC 2 data for international contexts. These choices highlight the necessity to also consider local overheating effects, particularly for buildings in larger cities, and the impacts of long-term climate change through climate projections.

This chapter delves into the complex interplay of factors fuelling this trend, including regional climate conditions, climatic changes, local building practices, and demographic shifts. It aims to unpack these interactions to pinpoint environmental contributors to overheating risks, highlighting their varying impacts across European regions. The analysis begins with *macroscopic patterns* such as temperature variations, the frequency and intensity of heatwaves, and demographic trends. However, an exploration limited to the macro scale does not suffice. A detailed examination of mesoclimatic conditions uncovers how topography, vegetation, and land use contribute to the creation of *unique microclimates* within regions. For instance, the cooling effects of hills and bodies of water like lakes or forests are advantageous, whereas large cities in valley locations often suffer due to inadequate ventilation which hampers the dispersion of anthropogenic heat and urban pollutants. Urban environments, characterized by the *urban heat island (UHI)* effect, present unique challenges. Heat sources such as traffic and HVAC (Heating, Ventilation, and Air Conditionings system) waste heat, along with structural elements like heat-retaining urban masses and dark surfaces of roofs and streets, exacerbate the urban climate in summer. The scarcity of vegetation and water bodies further intensifies these conditions.

2 Summer Conditions in Europe

2.1 Macroclimatic Air Temperature Conditions

A primary concern in the overheating risk assessment is understanding the **climatic indicators** that define the **European summer**. These indicators include the frequency of hot days, length of summer periods, temperature extremes and much more. The number of hot days, typically defined as days with maximum air temperatures exceeding 30 °C resp. 35 °C, varies significantly across Europe. Southern regions, influenced by a Mediterranean climate, often experience longer and more intense summer periods compared to the cooler, more temperate northern zones. This **number of hot days in Europe**, as depicted in the EEA- data in Figure 1 (left side), reveals a noteworthy pattern, particularly across the Iberian Peninsula, encompassing the southern regions of **Spain and Portugal**. These areas experience **more than 30 hot days annually** with temperatures soaring above 35 °C. However, when considering the capacity for **nocturnal cooling** — that is, tropical nights where temperatures do not drop below 20 °C coupled with days with temperatures over 30 °C— a slightly altered picture emerges (see Figure 1, right side). In this scenario, extensive areas of **Italy** also display significant heat stress, as do parts of **mainland Greece, Sicily, Corsica, and Cyprus**.

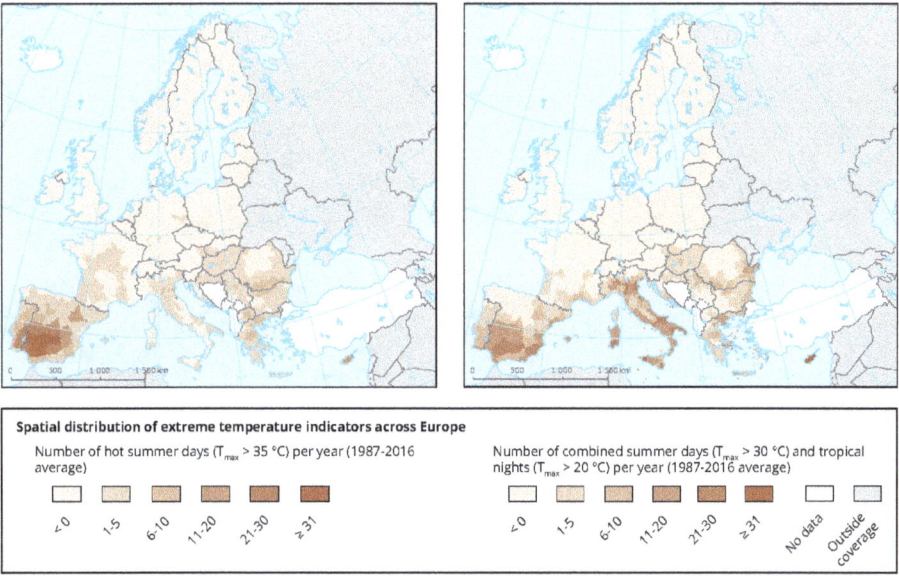

Spatial distribution of extreme temperature indicators across Europe

Number of hot summer days (T_{max} > 35 °C) per year (1987-2016 average)

<0   1-5   6-10   11-20   21-30   ≥ 31

Number of combined summer days (T_{max} > 30 °C) and tropical nights (T_{max} > 20 °C) per year (1987-2016 average)

<0   1-5   6-10   11-20   21-30   ≥ 31   No data   Outside coverage

Fig. 2.1: Spatial distribution of extreme temperatures: hot days (left figure) and combined hot days with tropical nights (right figure) across Europe according to long-term evaluations of the European Environment Agency (EEA) (1).

This adjustment in perspective underscores the ***importance of nighttime temperatures in assessing indoor heat stress***. While daytime heat is a recognized critical factor, the lack of sufficient cooling at night can exacerbate the impact on human indoor comfort and health, as well as on the built environment.

Meteorological patterns causing these temperature variations within Europe are shaped by a complex interplay of factors, with ***latitude*** playing a central role in determining the ***influx of solar radiation***. The Earth's tilt and its orbit around the sun mean that different latitudes receive varying intensities and durations of sunlight. While cities at similar latitudes should theoretically experience comparable solar radiation levels, ***other factors*** significantly influence local weather conditions as well. These include ***proximity to water bodies*** and topographical features.

From the temperate oceanic more humid climate in the west to the more continental and dry climates in the east of Europe, the variation impacts the intensity of summer heat. ***Continental climates*** exhibit ***larger diurnal and yearly temperature variations***, whereas ***maritime climates experience milder temperature fluctuations***. Figure 2.2 illustrates the impact of proximity to significant water bodies using the examples of weather stations near Warsaw, Hannover, and Amsterdam during July 2023. It's important to note that none of these cities are significantly influenced by other geographical features such as mountain ranges, or extensive forested areas.

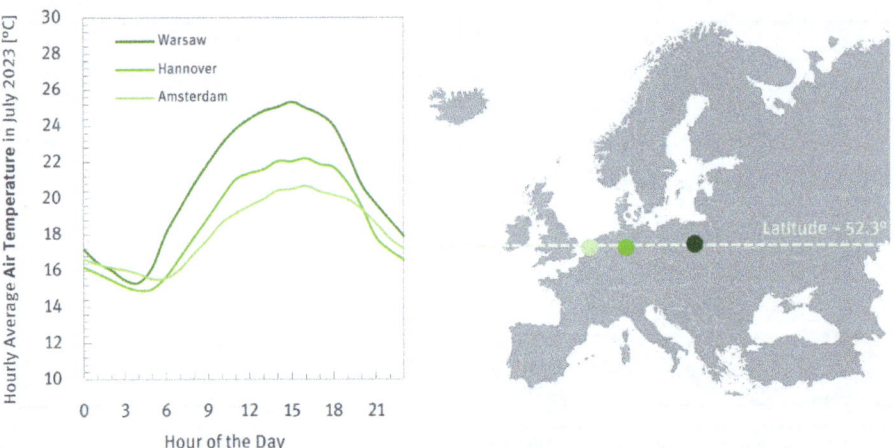

Fig. 2.2: Comparison of three different European locations in their average daily air temperature profile in July 2023. They are all approximately situated on the same longitude of 52.3° N. Hourly datasets obtained via NOAA (2).

Warsaw, Hannover, and Amsterdam, despite being situated at ***nearly identical latitudes*** of 52.3° N, exhibit notable ***differences*** in their temperature patterns, especially in their ***diurnal temperature ranges***. Interestingly, these cities achieve ***comparable*** *nighttime temperatures*, indicating a consistent nocturnal cooling pattern. However, the variation in their daily temperature peaks can be primarily attributed to their differing proximities to large bodies of water due to lower heating of air temperatures above water surfaces and higher degree of cloudiness for marine climates. *Warsaw*, located near the Baltic Sea but surrounded more closely by land masses, experiences a distinct *continental climate* characterized by pronounced daytime temperature rises and higher maximum air temperatures in the midday. In contrast, *Hannover*, positioned at a moderate distance from the North Sea and the Baltic Sea, displays a milder climate with less pronounced *temperature fluctuations*. *Amsterdam*, influenced significantly by the nearby North Sea with peak summer water temperatures around 20 °C, experiences a cooling effect during the day due to *maritime influences* moderating extreme temperature swings. These maritime effects include not only the stabilizing presence of water masses with relatively constant temperatures but also higher air humidity and increased cloudiness due to this humidity. However, during the night, Amsterdam's temperatures align more closely with those of Warsaw and Hannover.

2.2 Macroclimatic Solar Radiation Patterns

Global solar radiation, pivotal for overheating risk appraisal, demonstrates significant geographical variance. This is prominently illustrated in the enhanced solar irradiance across the Mediterranean in Figure 2.3 (right map), attributable to its higher latitudinal positioning, which sharply contrasts with the lower radiation intensities in the less sun-drenched climes of Northern Europe. Global radiation consists of *direct and diffuse components*. Direct radiation is the unobstructed solar energy that strikes the Earth surface, whereas diffuse radiation is scattered from the vast sky dome, moulded by atmospheric elements like clouds, molecules, and aerosols. *Visibility*, serving as an indirect *measure of solar radiation's direct-to-diffuse ratio*, intimates that the Mediterranean's clear, expansive skies predominantly facilitate direct sunlight, whereas the *higher latitude regions exhibit a dominance of diffuse radiation* due to lower visibility, often a by-product of a higher cloud coverage. Visibility and aerosol content in the atmosphere are intricately linked. Aerosols, those fine particulates suspended in the atmosphere, originate from varied sources such as natural sea sprays to urban pollution. They act as the substrates for cloud droplet formation and directly alter the solar radiation balance by scattering and absorbing sunlight, thus increasing the proportion of diffuse radiation. Higher aerosol concentrations typically result in lower visibility. For instance, in urban areas where aerosol emissions from vehicles and industry are higher, visibility tends to be reduced compared to cleaner rural areas.

Topographical influences are profound, as evidenced by *Europe's major mountain ranges*. The imposing Alps, Pyrenees, and Carpathians, with their significant elevation, ***foster cloud accumulation*** and, consequently, increased diffuse radiation due to orographic lift on their windward sides. In Northern-European ***coastal zones*** such as the coast of Brittany in France or the German North Sea coast, the nuanced relationship between geographic location, atmospheric conditions, and the solar irradiance received at Earth's surface is particularly pronounced. These areas often experience an uptick in cloud cover and a corresponding increase in diffuse radiation, as evidenced in Figure 2.3. This pattern contrasts with Southern Europe, where clearer skies over the Mediterranean result in higher solar radiation inputs and fewer cloudy days. The aggregated effect of these factors yields a complex and diverse solar radiation distribution across Europe.

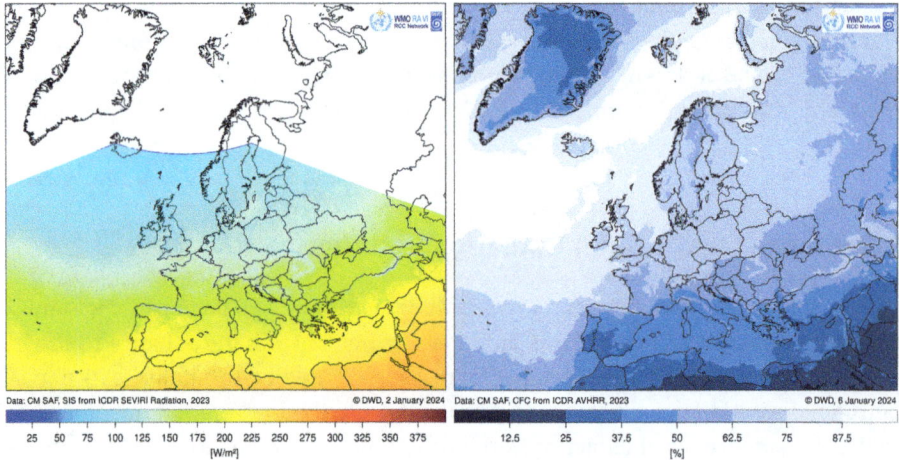

Fig. 2.3: Distribution of cloud fractional cover (right graph) and distribution of average global radiation (left graph) in 2023. Summarized figure according to DWD.

The ***impact of solar radiation on outdoor temperature levels*** can be vividly illustrated by examining the case of three European cities in Figure 2.3: Oslo, Warsaw, and Rome. These cities, chosen for their distinct latitudinal positions, offer a clear perspective on how global solar radiation influences local temperature regimes. In July 2023, average daily profiles of global solar radiation and temperatures were analysed for these locations. The mean daily global radiation, as depicted in Figure 2.3, indicates a decrease in solar radiation with increasing latitude. Oslo, located at a latitude of approximately 59.9° N, exhibits maximum solar radiation values of around 540 W/m² on a horizontal surface. Conversely, Rome, situated at a lower latitude of about 41.9° N, demonstrates significantly higher horizontal radiation values, approximately 800 W/m². This ***variation in solar radiation*** is a primary ***driver for the differences in*** mean ***air temperatures*** observed at these sites. In July 2023, the average temperature in Oslo was roughly 11 degrees Celsius lower than that of Rome. This disparity highlights the profound impact of solar radiation on the thermal environment of a location.

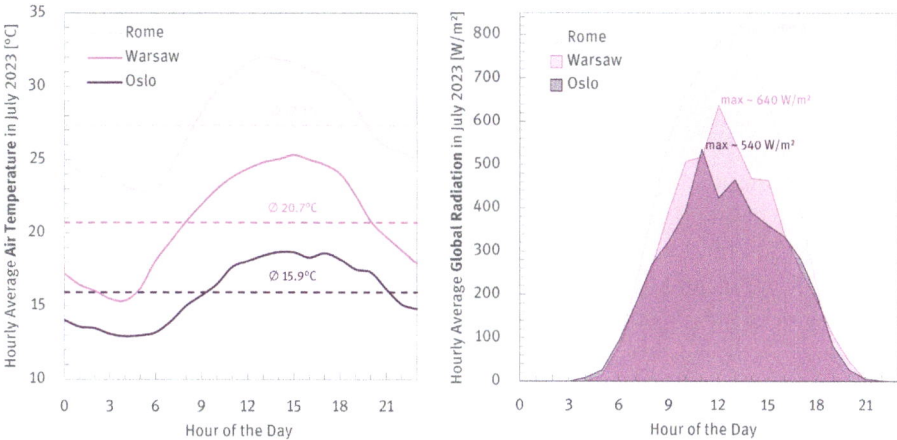

Fig. 2.4: Average diurnal temperature and global solar radiation variation in July 2023 at three different locations: Rome (~42° N, ~ sea level), Warsaw (~52° N, ~100 m above sea level) and Oslo (~60° N, ~ sea level). Hourly datasets obtained via NOAA (2).

While temperature and solar radiation serves as key indicators, they do not offer a complete depiction of summer thermal comfort. ***Other factors*** such as ***relative humidity and wind patterns*** also play critical roles. High humidity, often accompanying heatwaves, can intensify the perceived temperature, and consequently affect comfort levels.

2.3 Trends in Macroclimatic Weather Conditions

In recent years, **Western Europe** has experienced an ***increase in the frequency and intensity of heatwaves***. A compelling study (3) reveals that, from 2018 on, heat waves in Europe have increased three to four times faster than in other northern mid-latitudes, such as the United States and Canada. Over the last four decades, these scientists have observed that the jet stream — high-speed winds that circulate at altitudes of 5 to 10 km across Eurasia — has undergone modifications that correlate strongly with the increased frequency and intensity of heatwaves. One key phenomenon identified is the emergence of **double jet states**, where the jet stream splits into two distinct branches. The study highlights that while the total number of double jet events per year has remained fairly constant, their **duration has increased**, making them more likely to contribute to prolonged periods of extreme heat. This change accounts for about 30% of the heatwave trends across Europe and nearly all the trends in

Western Europe. This is in contrast to regions like the Mediterranean and Eastern Europe, where factors such as dry soils play a more significant role in heatwave development (3). This increase in frequency and intensity of heatwaves in Western Europe, can be further contextualized with the events of *July 2023*. Europe, specifically countries like Italy, Spain, France, Germany, and Poland, experienced a major heatwave, with forecasted air temperatures in Sicily and Sardinia potentially reaching up to 48 °C. This could mark some of the **hottest temperatures ever recorded in Europe**. The data from the Copernicus' radiometer instrument on 17 July 2023 showed land surface temperatures of 45 °C in Bucharest and Rome, and even higher in Catania and Nicosia, reaching 50 °C (4). It is important to note that air temperatures and land surface temperatures are not identical, although land surface temperatures can serve as a reliable indicator of extreme heat events.

2.3.1 Demographic Trends in Europe

In examining the risks associated with heat-related phenomena in Europe, such as increased mortality, morbidity and productivity losses, demographic structure emerges as a critical factor. **Older populations** are generally more **vulnerable to heat** (see next Chapter), with countries like Italy and Germany showcasing this trend within Europe. Both these nations, with average ages exceeding 45 years, are among the older populations in Europe. However, the pace of aging differs, with Italy experiencing more rapid demographic shifts compared to Germany (5).

The **internal demographic landscapes** within these countries reveal significant regional disparities. In Germany, there is a pronounced East-West gradient, with the eastern regions typically having a higher average age, often exceeding 50 years. Italy, meanwhile, displays a notable North-South divide, where the northern regions exhibit greater signs of aging compared to the south (6). Other Southern European countries, including Greece, Spain, and Portugal, have also seen their average ages surpass 45 years, with similar rates of increase to Italy (5,6). This aging trend, coupled with the increasing incidence of intense summer heatwaves, creates a problematic scenario. **Regions most impacted by this convergence of aging populations and heightened heatwaves** include **Northwest Spain and Eastern Portugal, Northern Italy, widespread areas of Greece, and parts of Bulgaria, Romania, and Southwest France**. These areas record heat-related mortality rates exceeding 3 deaths per thousand inhabitants, underscoring the significant health impacts of rising temperatures on aging communities (7,8).

Thus, Europe faces a dual challenge: demographic shift towards older populations and increasingly severe heatwaves, particularly in regions where these two phenomena intersect most acutely.

2.4 Mesoclimatic Conditions

The mesoclimate, which bridges the gap between the microscale of individual buildings or small landscapes and the macroscale of large geographical regions, plays a critical role in shaping summer conditions in various environments. Its influence is seen in the interaction between natural landscapes, such as forests and fields, and human-modified areas, including urban settings and agricultural lands.

Forests and fields significantly modulate the local climate. Forests, with their high density of trees and vegetation, act as natural air conditioners. They provide shade, reducing ground-level temperatures, and their transpiration process releases moisture, contributing to cooler and more humid air. *Forests* show strong interactions and strongly altered climate conditions at their edges if they are close to urban areas (9). This effect is particularly noticeable during the summer months when the temperature difference between forested and non-forested areas can be substantial. In contrast, open fields, especially those with low vegetation or crops, may have less of a cooling effect, sometimes contributing to higher local temperatures due to direct sun exposure and less moisture release.

The presence of *water bodies* like rivers, especially in valley regions, or large lakes, also significantly influences the mesoclimatic. These water bodies often create localized microclimates characterized by lower temperatures and higher humidity levels. Rivers in valleys can lead to the formation of cool air pools under stable atmospheric conditions, providing relief during hot summer days. Large lakes act as thermal buffers, absorbing heat during the day and releasing it slowly, thus moderating the temperature variations between day and night.

Human-made factors profoundly impact the mesoclimatic. The degree of land sealing, characterized by the extent to which natural land is covered by impermeable materials like concrete, plays a significant role. High sealing rates lead to increased surface temperatures and reduced soil moisture, exacerbating the urban heat island effect. This effect is particularly pronounced during summer, when the accumulated heat from sun-exposed surfaces elevates the local temperature, often making urban areas significantly hotter than their rural surroundings.

Dense urban construction contributes to this phenomenon by trapping heat, reducing wind flow, and limiting the area available for vegetation. Furthermore, *anthropogenic heat sources* such as vehicles, industrial processes, and air conditioning systems release additional heat into the environment, further elevating urban temperatures. The impact of these heat sources varies over the course of a day and is most pronounced during the afternoon and early evening hours in summer.

2.5 Urban Climate and the Heat Island Effect

The ***Urban Heat Island (UHI)*** effect, a phenomenon where urban areas experience higher temperatures compared to their rural surroundings, can be quantified using various climatic differences (10). One of them is the surface temperature difference between cities and their surroundings, quantified as ***Surface Urban Heat Island (SUHI)*** effect. Mentaschi et al. evaluated the SUHI from 2003 to 2020, using high-resolution global datasets. ***Intra-city hotspots*** were identified where extreme ***SUHI is up to 10–15 K***. Especially temperate and humid regions are characterized by high SUHI values. The study shows an ***average worldwide increase*** in SUHI, attributed to urbanization, more frequent heatwaves, and Earth's greening (11).

Urban areas are also characterized by ***reduced wind speeds*** due to their unique aerodynamic shape. Tall buildings in cities act as obstacles, disrupting wind flow, which is therefore more pronounced in rural areas. This reduction in wind speed contributes to the stagnation of air and exacerbates the UHI effect.

Regarding ***humidity***, urban areas have a different profile compared to rural settings. While the absolute humidity, the total amount of water vapor present in the air, might not necessarily differ between urban and rural areas, the ***relative humidity*** in cities is always ***effectively lower***. This is because warmer urban air can hold more moisture than cooler rural air, leading to a perception of drier conditions in cities. The ***absence of sufficient vegetation and water bodies*** in urban settings might furthermore contribute to lower absolute humidity levels. Vegetation and water bodies are crucial for maintaining atmospheric moisture, primarily through transpiration and evaporation processes. Their scarcity in urban landscapes leads to a drier atmosphere, exacerbating the UHI effect.

Longwave radiation, or the heat emitted by the Earth's surface, is also affected by urban structures. Urban materials like concrete and asphalt have ***high heat capacities***, acting as reservoirs for heat energy. During the day, these surfaces absorb a significant amount of solar radiation, which is then slowly released at night, leading to higher nighttime temperatures in urban areas compared to rural surroundings. The phenomenon is ***more pronounced when skies are clear and wind speeds are low***, as these conditions favour the formation of an inversion layer that traps warm air near the surface.

In summary, the ***urban overwarming*** is marked by ***higher temperatures, altered wind patterns, and reduced humidity and higher longwave emissions*** in urban areas compared to their rural counterparts. These climatic differences are influenced by various factors including urban geometry, the properties of urban materials, and human activities like traffic- or heating-related heat and particle emissions.

2.5.1 Urban Temperature Conditions

Focusing on **Berlin, Germany,** with its approximately 3.6 million inhabitants, we observe specific manifestations of the UHI effect. Berlin's macroclimatic conditions, characterized by a **moderate continental climate** with cold winters and warm summers, are ideally suited for nocturnal cooling. However, the urban climatic influences significantly alter this potential. An analysis of **challenging summer days** for the course of the year 2019 in Figure 2.5 highlights the UHI effect's impact in Berlin. These days are defined by a combination of maximum temperatures above 30 °C (hot days according to the German Weather Service, DWD) and nocturnal temperatures above 20 °C (tropical nights according to DWD). An illustrative **comparison** between the suburb weather station in **Southwest Berlin** and the **downtown station at Alexanderplatz** in 2019 reveals a **striking difference**: the former recorded only two such phases, while the latter experienced seven prolonged phases, emphasizing the severity of the UHI effect in more densely populated areas.

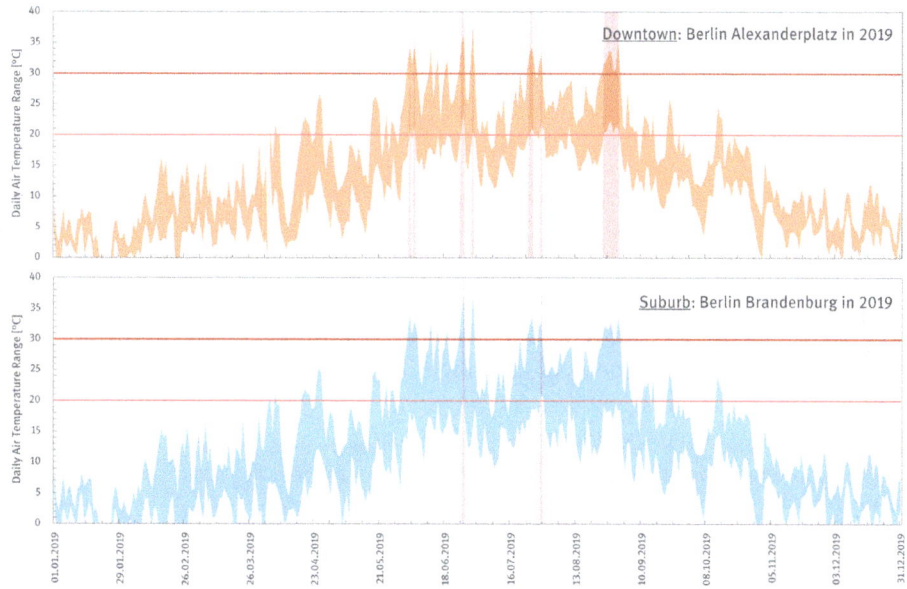

Fig. 2.5: Comparison of daily minimum and maximum air temperature in Berlin in 2018 for two weather stations, one in the city (Alexanderplatz, upper graph) and one in the suburbs (Berlin Brandenburg, bottom graph). Data obtained from (12).

Over the period 2016–2022, the **suburban station** Berlin Brandenburg recorded a sum of eight phases with these tropical night and hot day conditions, with a **maximum duration of three days in a row** as shown in the bar hight in Figure 2.6. In contrast, the **downtown station** recorded 25 such phases, some **lasting up to seven days consecutively**.

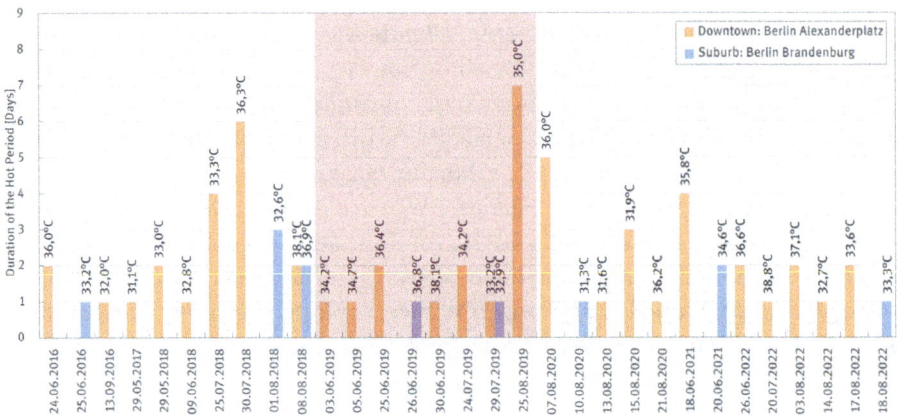

Fig. 2.6: Periods with a combination of hot days (max temperature above 30 °C) and tropical nights (temperature permanently above 20 °C) for Berlin downtown (orange) and suburban (blue) between 2016 and 2022. The red area highlights 2019, which is covered in Figure 2.5. Values within the bars (unit °C) show last day maximum temperatures of the phases. Data obtained from (12).

In the **case of the Berlin-Brandenburg** station, the year **2018** emerges as pivotal for summer assessment. This determination is based on the occurrence of the **longest heatwaves** in this year, characterized by two phases with a total of five burdensome days in 2018, compared to two phases with just two burdensome days in 2019. In contrast, for the **inner-city station at Alexanderplatz**, **both years, 2018 and 2019,** record 15 days meeting the criteria of high night temperatures (above 20 °C) and high maximum daily temperatures (exceeding 30 °C). However, the longest heatwave, spanning seven days, is recorded in 2019. Interestingly, for both the Berlin-Brandenburg and Alexanderplatz stations, the highest temperatures are recorded in the year of 2019. Suburban air temperatures raise up to 36.8 °C, while downtown temperatures raise up to 38.1 °C, both in June 2019.

 This **discrepancy in critical years between stations** underscores the **importance of localized climate data analysis in urban areas**. It highlights the challenge in selecting appropriate climatic datasets for indoor overheating assessment, as different locations within the same city can exhibit significantly different climatic behaviors. Nevertheless, under these considerations, the climatic conditions at Berlin-Alexanderplatz may be regarded as **analogous** to those experienced in the **Mediterranean region with typically more than 11 burdensome summer days per year**, as depicted on the European macroscopic map in Figure 2.1. This pattern highlights the significant physiological stress experienced by residents in urban areas during summer. Low **night temperatures** are essential for **restful sleep** and determine the **effectiveness of night-time ventilation for free cooling** of buildings. The lower the night temperatures, the greater the cooling potential of buildings by window ventilation through nocturnal reduction of stored heat. However, when night temperatures remain high, much higher air exchange rates are needed to utilize nocturnal cooling effectively.

The **duration of straining summer phases** is particularly **relevant for building design**. Buildings can only compensate for a limited duration of heat waves. If it's too warm at night, the building fails to cool down to the previous day's temperature levels, causing indoor temperatures to rise progressively day by day. Once a certain threshold of comfort or tolerance is reached, technical measures for temperature reduction become necessary in these buildings. Factors like **relative humidity**, typically lower in the city, and **average wind speed** also **influence the temperature sensation and cooling capacity**.

2.5.2 Urban Wind Conditions

Wind speed is primarily influenced by several environmental factors including altitude, proximity to the sea, and terrain topography. In urban settings, average wind speeds tend to be lower due to the obstructive nature of closely packed buildings that impede airflow. However, the unique architectural landscapes of cities can also lead to localized increases in wind velocity, particularly in narrow street canyons where wind is funnelled between high-rise structures. These variations underscore the complex dynamics of urban microclimates.

An analysis of **wind speed measurements from a suburban station in Dresden** (Dresden Klotzsche) and an **inner-city station** (Dresden Neustadt) over the period 2018-2022 provides valuable insights into this aspect. The averaged **daily profiles** derived from these measurements, as depicted in a comparative format in Figure X, reveal notable differences in the day profiles and the value range. The data indicates that the **suburban station**, particularly in summer, exhibits a **pronounced daily wind speed profile** with an average level of approximately 4 m/s. In contrast, this pattern is **barely discernible in the inner-city location**, where the average wind speed drops significantly below 1 m/s in summer. Although in this case, differences in wind speeds cannot be solely attributed to urban effects, as the suburban measurement station is situated at a higher elevation and is not nestled in a sheltered river valley like the urban station, such characteristics remain typical for urban environments A critical observation is therefore that **wind speeds are lowest when air exchange is most needed**, namely **during summer and at night**.

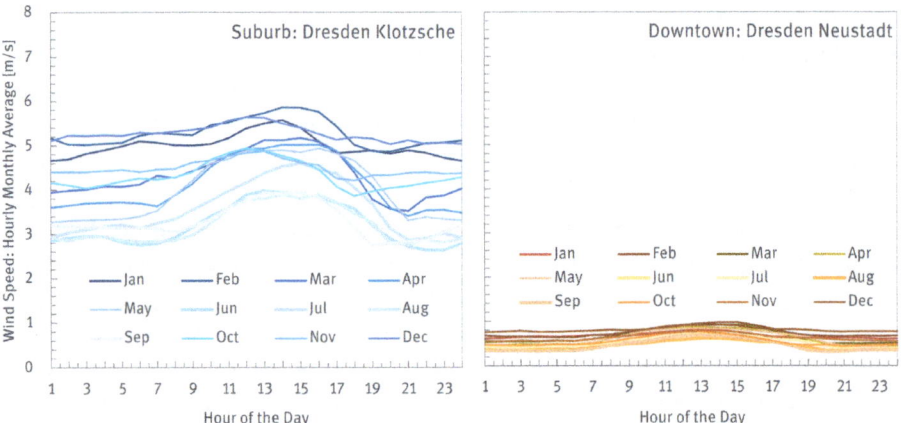

Fig. 2.7: Average hourly wind speed for each month in the period of 2020 to 2024 for the German city of Dresden with approximately 0.55 Mio inhabitants. Data obtained from (12).

Larsen's research further illuminates the numerical dependencies between temperature differences, air speeds and further parameters (13). It can be concluded from this research that such low air velocities lead to building *air exchange being predominantly driven by temperature differences* between indoor and outdoor environments. Given that nocturnal temperature differentials are likely to be minimal as well, it can be inferred from Larsen's measurement results that the *air exchange rate* of a single-sided ventilated building *during these summer phases* as recorded for Dresden would be limited to a maximum of about 1 h⁻¹. This value states that it takes one hour for an exchange of the entire room air volume, which is half of the night conditions assumed in the German overheating standard DIN 4108:2013 for residential buildings.

2.5.3 Urban Solar Radiation Conditions

In urban environments, the solar radiation conditions, including the balance between direct and diffuse radiation, exhibit distinct characteristics compared to surrounding rural areas. Typically, cities experience a *higher proportion of diffuse solar radiation* due to increased air pollution and particulate matter, which scatter sunlight. This scattering effect tends to increase the amount of diffuse radiation while simultaneously reducing the direct solar radiation that reaches urban surfaces. Furthermore, the *onset of diffuse radiation* in cities may *occur later* in the day as the sun rises above the horizon and urban structures, further impeded by potential smog or haze accumulation during early morning hours (14).

In contrast, ***direct solar radiation*** is ***generally lower*** in urban areas than in less obstructed rural environments, primarily because of the physical barriers presented by high-rise buildings and other urban infrastructure. These structures not only block and redirect sunlight but also create shaded areas that significantly alter the solar exposure patterns throughout the day. The variable reflectivity and emissivity of urban surfaces, such as those treated with specialized coatings or made of materials like low-emissivity glass, further modulate these patterns by altering the absorption and reflection of solar radiation. Such cool-colour material surfaces can effectively reduce heat absorption and control solar gain, contributing to the distinctive radiative balance in urban settings (14).

Moreover, ***urban aerosols*** play a crucial role in modifying the solar radiation received. Aerosols from combustion and industrial activities enhance the scattering of shortwave radiation, increasing the proportion of diffuse radiation over direct radiation. This effect is compounded by the vertical structure of the city, where high buildings and narrow street canyons can significantly alter the path and intensity of solar radiation, leading to complex patterns of shading and irradiance that vary throughout the day and across different surfaces (14,15).

2.5.4 Urban Microclimate

The concept of microclimate refers to the climatic conditions in a specific, localized area, often contrasting significantly with the broader regional climate. This scale typically spans small spatial areas, such as urban ***neighborhoods, parks, street canyons***, or even the ***immediate surroundings of a single building***. Microclimatic conditions are primarily determined by a variety of factors including vegetation, water bodies, and human-made structures. Additionally, the microclimate is affected by anthropogenic heat from vehicles, industrial processes, and air conditioning systems, which can significantly alter thermal conditions in localized areas.

The assessment of a building's risk of overheating due to microscale urban configurations necessitates ***detailed microscale urban climate simulations.*** These simulations are pivotal in capturing the nuanced interplay of local overheating, airflow dynamics, and diurnal meteorological cycles. ***Tools such as ENVI-met and PALM-4U*** offer a sophisticated three-dimensional representation of the urban environment, facilitating a comprehensive analysis by using meteorological time series data specific to a building's location. This method allows for the examination of how local open spaces influence indoor overheating. However, the approach is not without its limitations. The precision of these simulations depends heavily on the selection of accurate simulation boundary conditions and the construction of detailed city models, which require significant data and computational resources. These factors often restrict simulations to short spans, typically a few days during peak summer conditions.

3 Climate Data for Overheating Risk Assessment

3.1 Types of Climate Data Sets for Building Simulations

The selection of appropriate climate data sets is crucial for the architectural and structural design of buildings, particularly in the context of overheating risk assessment. There are primarily two types of climate data sets available: *measured and synthetic data sets*.

    Measured data can be obtained directly from the site or from official meteorological stations. These stations could be hosted or maintained by national weather services, environmental agencies, or other public authorities. In this case, they are highly reliable as they are typically built according to the requirements of the World Meteorological Organization (WMO) but can be limited in temporal and spatial coverage. Although data from unknown and uncertified sources can map the local climate at the building location, they harbor the risk of inaccurate measured values due to uncalibrated measuring equipment or inadequate maintenance of the measuring instruments. However, they can be a reasonable supplement wherever the requirements for 'undisturbed' climate data acquisition according to the WMO requirements are not met, e.g. for most urban locations. *High-quality measurement data* in accordance with WMO requirements can be obtained from numerous data sources, e.g.

- via the *Copernicus* program (*https://www.copernicus.eu/en/access-data*),
- via *freely available national data sets* like the German Meteorological Service DWD (*https://opendata.dwd.de/cf*) or
- via *service-based national data sets* like the British weather service MetOffice (*https://www.metoffice.gov.uk/*).

Synthetic data sets, on the other hand, are typically derived from these measurements. They utilize spatial interpolations and temporal extrapolations, building on trends found in the real data. In the context of building simulation for summer behaviour, synthetic datasets derived from climate models are crucial for forecasting future conditions in Europe. These models incorporate various climate scenarios, such as those from the IPCC (Intergovernmental Panel on Climate Change) to project changes in environmental variables relevant to building performance. These models feature various Representative Concentration Pathways (RCPs) which reflect a range of future climate conditions based on different emissions and socio-economic scenarios. These RCPs offer a spectrum of potential future climates, each corresponding to different levels of greenhouse gas concentration forecasts. The *accuracy of these models*, especially for application in Europe, has been a subject of extensive research. One key aspect is the process-informed bias correction, which improves the reliability of climate simulations. For instance, Maraun et al. (16) emphasized the significance of such correctional approaches in refining climate model outputs.

However, the reliability of specific models can vary significantly based on the region and the climatic variables under consideration.

Moreover, in the context of **multi-model assessments**, the **performance and interdependence of different models are considered**. Knutti et al. (17) introduced a climate model projection weighting scheme, which accounts for these factors. This approach underscores the importance of considering both the skill and the independence of models in ensemble studies, ensuring a more robust and reliable prediction of future climatic conditions.

3.2 Selection Criteria

In the domain of building assessment, particularly for assessing summer behaviour, the selection of representative climate datasets from extensive historical records is a critical step. This representativeness might aim to reflect **typical long-term climatic conditions** or to specifically represent a **typical hot summer scenario**. The selection process is non-trivial due to the multiple factors that influence summer conditions especially for the indoor evaluation context, making it a complex and multi-dimensional decision.

When choosing datasets for building simulation, relying solely on outdoor air **temperature data** will likely inadequately represent the risk of overheating. Similarly, basing the selection only on **solar radiation data** might also provide an incomplete picture. The complex interplay of various weather elements, such as air temperature, humidity, solar radiation, and wind speed, collectively influences the indoor environment of a building. The **individual impact of these weather elements** also varies significantly depending on the specific characteristics of the building or room in question. For instance, buildings with high air exchange rates are more sensitive to outdoor air temperature variations, while buildings with large glass areas are more affected by solar radiation. This distinction is crucial in accurately simulating the building's response to summer conditions.

The selection of appropriate climate data for building simulation, particularly focusing on the external air temperature element, involves the consideration of various annual metrics. Options for such **metrics** include **singular values** like maximum air temperature, **frequency distribution measures** like number of summer days or tropical nights, and **integral indicators** like the sum products for exceeding selected threshold temperatures.

A **comparison of different temperature and radiation metrics** is illustrated as follows in Figure 3.1 for the example of Dresden, a German city in the east of Germany, in the measurement period 1982 to 2022 for the air temperature and 1997 to 2023 for the global radiation, both at DWD weather station Dresden-Klotzsche.

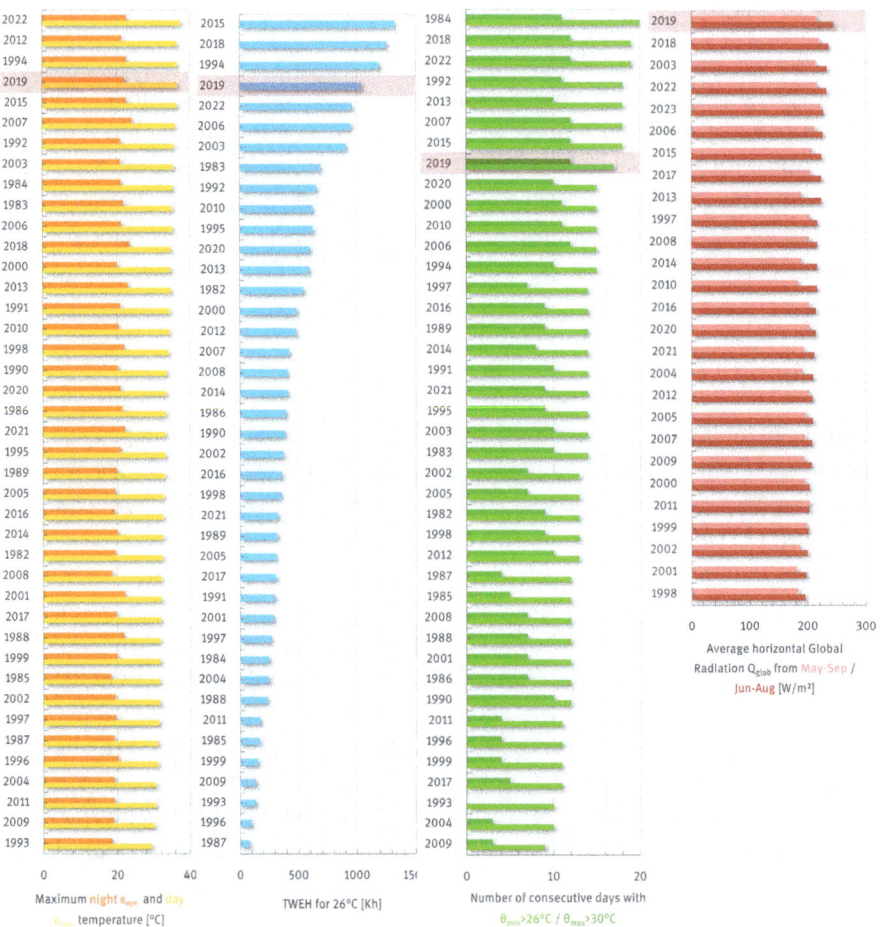

Fig. 3.1: Ranking of years according to different selected comparison metrics for Dresden-Klotzsche. Highlighted in red is the example of 2019, showing different ranking positions for different criteria. TWEH represents Temperature-Weighted Exceeding Hours. Data obtained from (12).

It reveals notable variations within a single parameter, such as maximum **temperature**, depending on the selected time frame. For instance, maximum temperatures during night and day show different rankings of years (as shown in the left column of Figure 3.1). This variation becomes even more evident when comparing rankings across different selection metrics like the integral value of the Temperature-Weighted Exceedance Hours (TWEH) for 26 °C (second graph from the left in Figure 3.1) or the number of consecutive days with a maximum temperature of 26 °C resp. 30 °C (second graph from the right in Figure 3.1). TWEH, as outlined by ASHRAE 55 and DIN 4108-2 (18,19) accounts for the magnitude and duration of temperature exceedances above a set threshold. A TWEH value over 1200 Kh/a for indoor air temperature is considered critically high for positive operational assessment (OA). This is the case for four years, although the displayed value is the outdoor air temperature. It becomes evident that different years yield varying levels of criticality based on the chosen parameters.

In the case of the presented temperature metrics, the year 2019, for example, might not be deemed the most critical. However, when considering **global radiation values**, the perspective shifts significantly. For the year 2019, the highest values of global radiation are observed during the summer months. This suggests that, in terms of solar radiation, 2019 could be considered more critical for building performance, especially in scenarios where solar gain plays a significant role, such as in buildings with large glass areas or limited shading. This highlights the **importance of solar radiation as a crucial factor in assessing building performance during summer**.

Interestingly, also the ranking of the years according to global radiation data changes when the **reference period** is altered. While 2019 shows the highest global radiation values in the summer months (June, July, August), its ranking changes if the period is extended to include May through September. It illustrates that the selection of the assessment period is a key factor in climate data analysis for building simulation, influencing the outcome of the ranking and, consequently, the interpretation of which year could be considered most critical.

To reconcile these variations, a **unified approach** could be summing up the ranking of different years based on the chosen evaluation criteria. However, this harbors the risk that small differences in values will have an excessive influence on the ranking result. Another method could involve a proportional averaging of normalized individual metrics. Both approaches aim to provide a comprehensive overview that accounts for the multidimensional nature of climate data and its impact on building performance.

3.3 Available Climate Data Sets

The precision in assessing indoor overheating risk of buildings crucially depends on the utilization of **representative data sets** in simplified models or building performance simulation (BPS). These data sets, in contrast to more comprehensive meteorological datasets provided by authorities or other sources, are refined to include only case-focused essential weather elements for a selected representative year. Like the requirements for winter and annual measurements, the summer case necessitates data sets encompassing air temperature and short-wave radiation as primary elements, alongside air humidity. Additional weather elements, though optional, can enhance the model's accuracy. These include long-wave sky (counter) radiation, wind conditions (direction and speed) or ambient illumination. Notably, the last two elements exhibit significant dynamism over time, rendering hourly values occasionally unsuitable. For most weather elements, however, an hourly temporal resolution, spanning precisely one year, has proven effective.

Further research suggests that while representative data sets are invaluable for BPS, the selection of these data sets should **consider both current climate patterns and future climate projections**. This consideration is essential to ensure buildings are designed to be resilient not just under current conditions but also under changing climate scenarios. Consequently, different data sets exist that either reflect current climate conditions or include future climate projections. Three examples include:

- The **'Test Reference Year' (TRY)** from the **DWD** for Germany in a former version from 2010 (based on measured data from for 15 regions of Germany) and the current version from 2017 (simulated meteorological data for a spatial grid with a resolution of 1 km² for Germany, based on measurements from 1995 to 2012 (20). One challenge in the compilation of these TRY datasets was the lack of solar shortwave radiation data the nearest radiation measurement stations exceed partially 60 km. This complicated the accurate interpolation of shortwave radiation parameters due to potential inconsistencies like varying cloud cover. To address these challenges, TRYs incorporate not only ground-based measurements of solar radiation, both diffuse and global, but also satellite data. Further, the interpolation process for wind and other meteorological parameters is meticulously detailed to account for variations caused by local geographical features like mountains and valleys. Amongst these features, these data sets contain a modified UHI modelling approach that estimates the influence of the hourly temperature increase using land use parameters instead of population data, combined with wind and cloud cover data. However, this approach has only been validated for two climate locations (Berlin, Potsdam) and shows an underestimation of the UHI offset in the middle of the day (17,18). Additionally, future projection years for the TRY are delineated, with datasets targeting projections for 2035 and newly updated datasets extending to 2045. The location-specific Test Reference Years (TRY) for 2045 are formulated based on projections of climate development up to this year, utilizing high-resolution regional climate models (RCMs) driven by global climate models (GCMs). These projections incorporate two distinct global greenhouse gas emission scenarios—moderate and high emissions—outlined in the Representative Concentration Pathways (RCPs) (20).
- The **'International Weather for Energy Calculations'** version 2.0 **(IWEC 2)** from **ASHRAE** offers meteorological data as 'typical' weather files for BPS for more than 3000 locations in the world. The data is derived from the Integrated Surface Hourly (ISH) weather data, originally archived at the National Climatic Data Center. IWEC 2 files are supported by a range of simulation tools (21).
- The **'Design Summer Year' (DSY)** weather files, specifically the DSY1 variant, are crucial for conducting overheating analysis in the UK, particularly in line with CIBSE TM59 guidelines. The DSY1 files represent weather data from a moderately warm summer and are utilized for operational analysis alongside measured meteorological data. CIBSE has updated the DSY files in collaboration with UK institutions like Exeter University and Arup, enhancing the accuracy of these datasets

to reflect current and future climate conditions. For instance, the Design Summer Year for London, detailed in CIBSE's TM49, is specifically tailored to assess and address the cooling needs during London's summers, considering the urban heat island effect and other local climatic influences (CIBSE TM59).

4 Climate Data's Impact on Indoor Overheating Risk

To analyse the ***impact of different standardised meteorological inputs on indoor comfort*** different representative climate data sets are used as input for the overheating risk assessment via building performance simulation ***for a residential building in Dresden, Germany***. Building performance simulations (BPS) were conducted using IDA ICE 4.8 software (22). The simulations utilized a consistent model of a Gründerzeit apartment in a block setting, adhering to the boundary conditions detailed in (23). The variables manipulated in these simulations included 13 different meteorological datasets and variations in location characteristics (city or suburb). This study focused on an attic bedroom on the Eastern façade, identified as having a high risk of overheating. Overheating intensity was measured using temperature-weighted exceedance hours (TWEH).

4.1 Indoor Overheating Risk based on Measured Climate

This analysis initially focused on assessing the impact of varying summer conditions on indoor overheating based on the time span 1991 to 2020, using recorded data from the Dresden-Klotzsche meteorological station, obtained from the DWD (12). The selected years—1996, 2007, and 2019—correspond to the coolest, average, and hottest summers, respectively, within this timeframe. The reliance on outdoor air temperature alone for this analysis simplifies the complex interplay of meteorological variables, such as solar radiation, wind, and humidity, which also critically affect indoor thermal comfort as shown in the previous section in Figure 3.1. The temperature analysis was therefore combined with global radiation parameters, as discussed in the previous section.

To assess the effect of the annual data sets used on the indoor overheating risk, the aforementioned room was evaluated with different limit temperatures with regard to the expected TWEH. These are shown in Figure 4.1 for each of the three cases analyzed, namely:

- ***Dresden***- Klotzsche ***1996 (case a) as the coolest year*** from 1991 to 2020 according to temperature conditions (number of hot days, tropical nights) and global radiation data measured at the DWD airport weather station.
- ***Dresden***- Klotzsche ***2007 (case b) as the average year*** from 1991 to 2020 according to the same selection criteria and based on the same station data.
- ***Dresden*** Klotzsche ***2019 (case c) as the hottest year*** from 1991 to 2020 according to the same selection criteria and based on the same station data.

The colored in Figure 4.1 bars indicate the TWEH parameters for individual temperature ranges and for the entire year, while the parameters next to the bars represent the THW parameters for the limit temperature of 26 °C and for the summer period from June to August. The comparison demonstrates significant annual variations in overheating intensity. For example, the TWEH above 26 °C was only 270 Kh/a during the coolest summer, while the hottest summer experienced TWEH exceeding 3000 Kh/a, highlighting substantial annual variability in indoor comfort due to natural summer condition fluctuations.

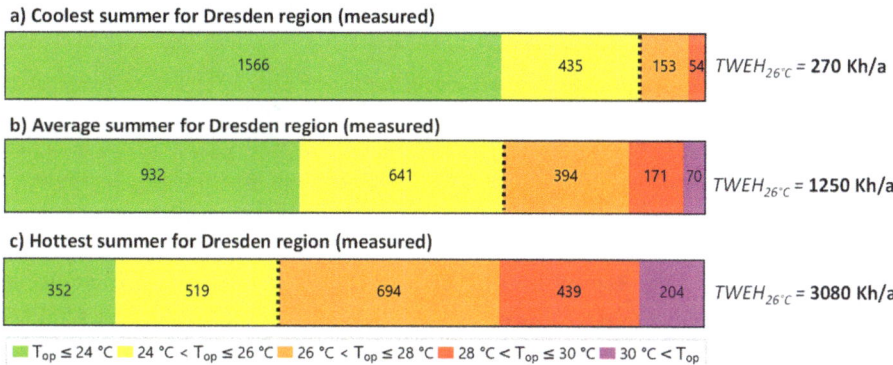

Fig. 4.1: Results of the simulated indoor air exceedance hours for 1996 (a: coolest), 2007 (b: average) and 2019 (c: hottest) in Dresden-Klotzsche. The numbers in the bars represent hours of exceedance per year within each temperature range and the numbers beside the bars represent the TWEH values (including exceedance intensity and duration) spanning June, July and August.

These results underscore the inadequacy of using an average summer for operational assessment in building performance simulations (BPS), particularly in addressing the heat resilience of buildings during unusually hot summers. Therefore, it is advisable to incorporate data from warmer than average, and possibly extreme, summers to ensure buildings remain comfortable under challenging conditions. The Test Reference Year (TRY) data sets from the German Weather Service (DWD) offer therefore extreme years which are designed to represent summers warmer than the average over recent decades, providing a more robust foundation.

4.2 Indoor Overheating Risk based on Reference Data Sets

In assessing indoor overheating through Building Performance Simulation (BPS), various reference climate data sets are used to represent typical meteorological conditions. These include both measured data, such as TRY2010, and a combination of measured and simulated data, like TRY2017. They are designed to reflect representative summer, winter or yearly weather patterns. This study includes a combination of typical annual conditions (TRY 2010) and a non-extreme UHI-effect offset (TRY 2017).

Several national and international agencies offer these meteorological data sets, each incorporating different regional climatic nuances and varying significantly in temporal basis and spatial resolution of the underlying meteorological data. Four distinct climate data sets were juxtaposed:

- *TRY2017 for the city of Dresden (case a)* - Features enhanced spatial resolution, enabling site-specific analysis through advanced software applications provided by the DWD (German Weather Service).
- *TRY2010 for Region 12 (Mannheim,* current reference location for Dresden in the national standard DIN 4108-2:2013*) (case b)* - Reflects older standards where datasets were created for broader geographic zones across Germany. Zone Mannheim is the hottest year among these 16 TRY 2010 data sets.
- *TRY2010 for Region 4 (Potsdam,* outdated reference location for Dresden in the previous version of the map referenced in DIN 4108-2:2013*) (case c)* - Reflects older standards where datasets were created for broader geographic zones across Germany. Zone Potsdam represents a medium-hot TRY.
- *ASHRAE IWEC 2 for Region Dresden (case d)* - Provides an intermediate level of assessment, not tailored as specifically as the newer TRY2017 but more detailed than older TRY versions.

The choice of data set significantly influences the Temperature-Weighted Exceedance Hours (TWEH) results as shown in Figure 4.2:

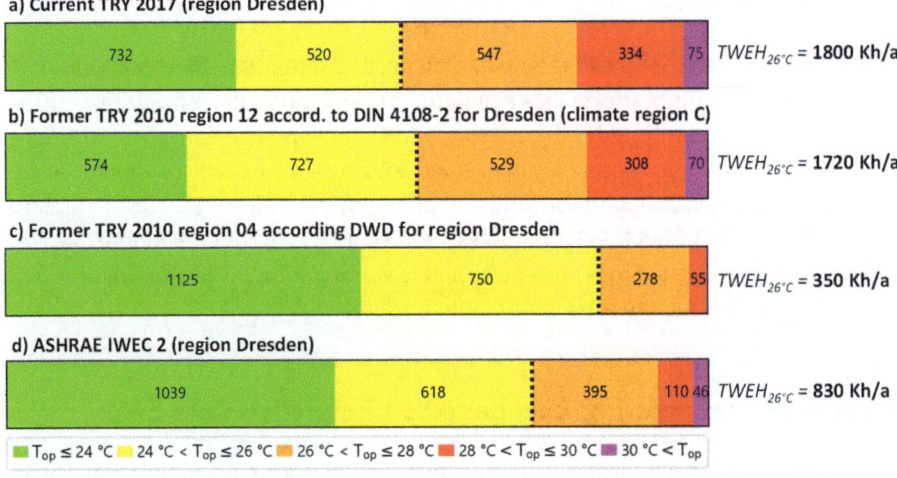

Fig. 4.2: Results of the simulated indoor air exceedance hours for Reference Years according to the DWD for the recent data sets TRY 2017 for Dresden (a) for previous TRY 2010 region 12 (b) and region 4 (c) and the IWEC-2 data set for Dresden. The numbers in the bars represent hours of exceedance per year within each temperature range and the numbers beside the bars represent the TWEH values (including exceedance intensity and duration) spanning June, July and August.

The outcomes for TWEH for a threshold temperature of 26 °C are ranging from a low of 350 Kh/a to a high of 1800 Kh/a. Such variation can drastically affect whether a room's heating condition is deemed adequate or not. According to the German stand-ard DIN 4108-2, a critical threshold is set at 1200 Kh/a. The TRY2010 for region 12 (Mannheim) and the high-resolution TRY2017 suggest the highest overheating inten-sity for Dresden, with TRY2017 including local Urban Heat Island (UHI) effects.   In contrast, the older TRY2010 assuming region 4 (Potsdam) results in much lower, non-critical overheating levels. The ASHRAE IWEC 2 data set results in an intermediate overheating assessment.

This significant contrast in TWEH outcomes, varying by a factor of four, prompts a critical evaluation of which standard year most accurately reflects the summer con-ditions at a specific location like Dresden. Comparing these to previously measured data from cool, average, and hot summers in Dresden between 1991 and 2020, where TWEH values were 270 Kh/a, 1250 Kh/a, and 3080 Kh/a respectively, it's clear that standard years can range widely.

The standardized years' TWEH for the coolest and hottest conditions generally match the observed data's range for past summers. However, the average summer TWEH of 1250 Kh/a indicates that both ASHRAE IWEC 2 and the older TRY2010 for region 4 likely represent cooler-than-average summers. Therefore, both the recent TRY2010 for region 12 and TRY2017 are potentially more appropriate for operational assessments in Dresden, as their TWEH values exceed the average observed TWEH.

4.3 Indoor Overheating Risk based on Future Data Sets

To effectively address the challenges posed by global warming in building perfor-mance simulations (BPS), an updated approach utilizing future-oriented meteorolog-ical data is essential. This ensures that buildings undergoing Overheating Analysis (OA) maintain acceptable overheating risk levels throughout their lifecycle, typically spanning at least 30 years. The following approaches were employed in the subse-quent study to assess the impacts of future climate scenarios on the overheating risk of buildings:

- *TRY2010- Projection 2035 (case a) for the Region 04* provided by the DWD for projections up to 2035, incorporates regional climate models covering the period from 2021 to 2050. This dataset is constructed for broader geographic zones across Germany.
- *TRY2017- Projection 2045 (case b)* for the city of Dresden, also from the DWD, offers more detailed projections up to 2045 with its 1 km² grid resolution. This allows for highly localized climate modelling.
- *A custom approach (case c)* uses climate projections resulting in a distribution of outdoor air temperature increases for future summers. This was assessed by comparing to historical data from 1991 to 2020 at the DWD station Dresden-

Klotzsche, identifying summer 2015 as the most representative of future average conditions for the period 2031-2050. Explanations are included in (23).

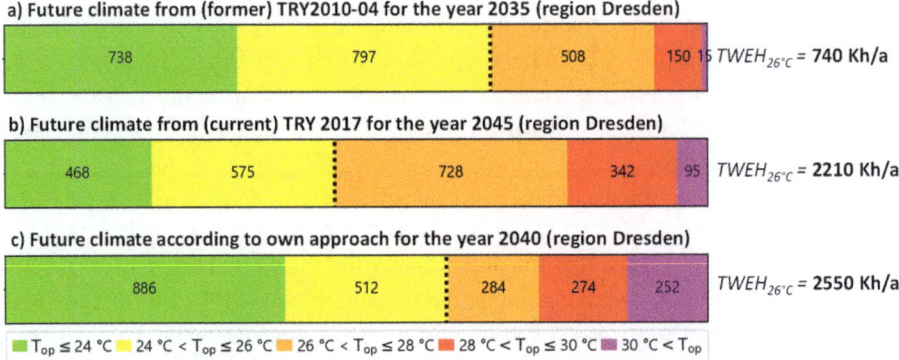

a) Future climate from (former) TRY2010-04 for the year 2035 (region Dresden)

| 738 | 797 | 508 | 150 15 | $TWEH_{26°C}$ = 740 Kh/a |

b) Future climate from (current) TRY 2017 for the year 2045 (region Dresden)

| 468 | 575 | 728 | 342 | 95 | $TWEH_{26°C}$ = 2210 Kh/a |

c) Future climate according to own approach for the year 2040 (region Dresden)

| 886 | 512 | 284 | 274 | 252 | $TWEH_{26°C}$ = 2550 Kh/a |

■ T_{op} ≤ 24 °C ■ 24 °C < T_{op} ≤ 26 °C ■ 26 °C < T_{op} ≤ 28 °C ■ 28 °C < T_{op} ≤ 30 °C ■ 30 °C < T_{op}

Fig. 4.3: Results of the simulated indoor air exceedance hours for future data sets according to the DWD for the TRY 2010 projected for the year of 2035 for Dresden (a) for TRY 2017 projected for the year of 2045 as well as for the custom case (c). The numbers in the bars represent hours of exceedance per year within each temperature range and the numbers beside the bars represent the TWEH values (including exceedance intensity and duration) spanning June, July and August.

The findings in Figure 4.3 illustrate significant variations in the Temperature-Weighted Exceedance Hours (TWEH) across these data sets. The analysis reveals that the future-oriented TRY2035 and TRY2045 datasets project considerably higher TWEH values compared to historical datasets, reflecting the potential increase in overheating risk due to global warming. The TRY2017, with its inclusion of the local Urban Heat Island (UHI) effect and high spatial resolution, shows the highest TWEH values, underscoring the importance of incorporating detailed local climatic effects in OA.

It's important to note that standard meteorological datasets like the older TRY2010, tailored for broader regional assessments, may not accurately reflect the localized climatic conditions essential for precise building performance evaluations. The use of high-resolution, localized data from newer models like TRY2017 provides a more realistic representation of future climate scenarios, crucial for designing buildings resilient to increased temperatures.

4.4 Indoor Overheating Risk and the UHI-Effect in Data Sets

In urban environments, particularly in free-running buildings that lack technical cooling systems, the intensity of overheating is heavily influenced by nocturnal outdoor temperatures, which are crucial for facilitating indoor cooling through natural ventilation. This phenomenon is markedly significant in areas with dense urban infrastructure and limited natural landscapes, where Urban Heat Island (UHI) effects elevate nighttime temperatures substantially compared to rural surroundings.

The city of Berlin, with a population of approximately 5 million inhabitants, serves as a significant case study due to its pronounced UHI effect, unlike Dresden, which has a comparatively smaller extension. Data from two distinct meteorological stations for the year 2019 were analysed to understand the impact of the Urban Heat Island (UHI) effect in Berlin. The results of this analysis are summarized in Figure 2.5 and 2.6. These two measured datasets were complemented by a modified dataset in which the UHI effect was imprinted following the previously mentioned approach of TRY2017 (20,24). This step aims to assess the representativeness of the generated urban climate temperature profile in comparison with the inner-city measurements. To encapsulate the UHI effect and its impact on urban warming, three specific datasets were employed:

– **Berlin-Brandenburg (BER)** Airport Station **(case a) 2019**: Situated in a suburban area, reflects less influence from the UHI, showcasing lower temperatures that are more representative of non-urban conditions.

– **Berlin-Alexanderplatz 2019 (case b)**: Located in the city center, this station provided key temperature readings that typically represent the heightened thermal conditions due to UHI effects in urban cores.

– **Modified Meteorological Dataset for BER 2019 (case c)**: This dataset, applied to the suburban station's temperature data, incorporates the UHI effect by adjusting the temperature profile to reflect urban warming conditions without direct measurements from the city centre.

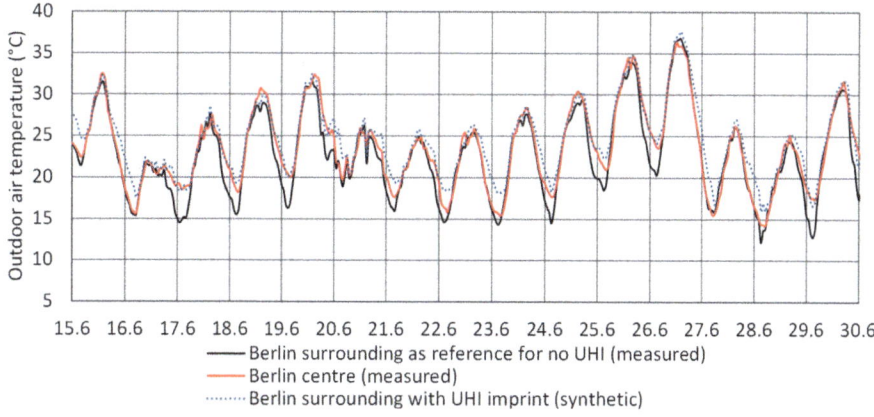

Fig. 4.4: Outdoor air temperature courses for two exemplary weeks in summer 2019 measured for Berlin centre (Alexanderplatz (12)) and Berlin surrounding (Berlin-Brandenburg(12)) pointing out the impact of local urban heat island effect. The dotted line represents the synthetic UHI imprint on the Berlin-Brandenburg station according to (24).

The data in Figure 4.4 illustrated notable temperature disparities between these locations and the modified data set, especially during nighttime, with urban areas showing temperatures 2-4 K higher on average, and at times up to 7 K higher than the suburban areas. To illustrate the impact of these three datasets, the overheating risk for the previously discussed application case was simulated. The results are summarized in Figure 4.5.

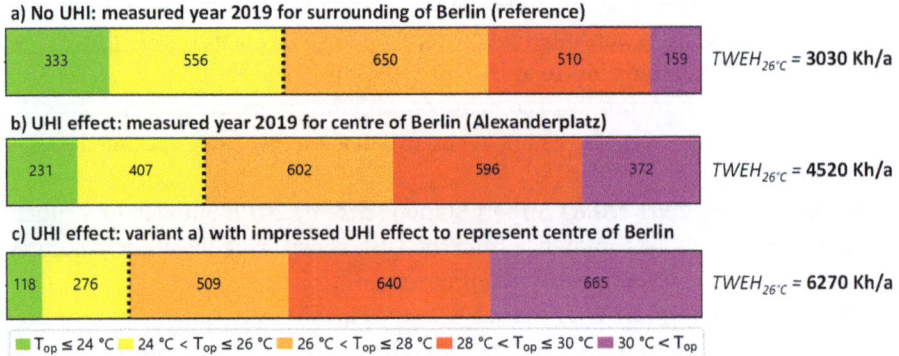

Fig. 4.5: Results of the simulated indoor air exceedance hours using urban datasets from DWD measurements for Berlin-Alexanderplatz (b) and Berlin-Brandenburg (a), along with a custom scenario (c). The numbers in the bars represent hours of exceedance per year within each temperature range and the numbers beside the bars represent the TWEH values (including exceedance intensity and duration) spanning June, July and August.

Figure 4.4, referenced in the study, vividly illustrates the effectiveness of this approach, showing a strong alignment between the synthetically adjusted temperatures and the actual observed temperatures in central Berlin, thereby validating the method's accuracy in replicating real-world UHI effects. This figure effectively underscores the critical need to consider UHI modifications in meteorological datasets used for building performance simulations (BPS) to ensure that overheating assessments reflect the true urban thermal environment.

4.5 Comparative Analysis of Meteorological Data Results

To synthesize the impact of different climatic datasets on simulation outcomes, the following overview (Figure 4.6) compares the characteristic values of all meteorological datasets. This comparison continues the analysis initiated in Section 3 regarding the selection of an appropriate dataset for modeling a critical summer. It becomes evident that the choice of dataset heavily depends on the criteria set for the simulations. For instance, selecting a dataset based on solar radiation inputs would yield a different choice compared to selecting based on the number of tropical nights. This variability highlights the complexity and critical importance of dataset selection criteria in accurately modeling and predicting summer overheating scenarios.

Category of comparison	Meteorological data set	Annual average outdoor temperature	Average summer* outdoor temperature	Maximum outdoor temperature	diffuse horiz. solar irradia-tion in summer*	direct horiz. solar irradia-tion in summer*	Numer of hot days summer*	Number of tropical nights in summer*	Number of warm nights in summer*
		°C	°C	°C	kWh/m²	kWh/m²	-	-	-
1. Deviations of annual summer conditions	coolest summer for Dresden region (measured), year 1996	7,1	16,5	31,2	220	203	2	1	1
	average summer for Dresden region (measured), year 2007	10,5	18,7	36,1	230	231	4	3	9
	hottest summer for Dresden region (measured), year 2019	11,2	20,9	36,7	223	324	17	6	21
2. Standardised meteorological input	former TRY2010-12 for Dresden region accord. DIN4108-2	9,5	18,1	35,4	220	245	5	1	8
	former TRY2010-04 for Dresden region accord. to DWD	11,1	19,5	36,3	229	256	11	2	10
	current TRY 2017 for Dresden centre	10,4	19,1	33,5	227	236	6	3	19
	ASHRAE IWEC 2 for Dresden region	8,9	17,7	33,2	254	188	3	2	9
3. Considering climate change	Year 2035 from former TRY2010-04 for Dresden region	10,7	19,2	35,4	222	266	9	1	9
	Year 2045 from current TRY2017 for Dresden centre	11,7	19,9	32,6	226	217	5	8	28
	Year 2040 by our approach for Dresden region	10,8	19,7	36,7	225	270	21	8	22
4. Considering urban heat island effect	Berlin surrounding - no UHI (2019, measured)	11,5	21,3	37,9	211	311	25	8	22
	Berlin centre (2019, measured)	12,5	22,2	38,1	211	311	30	22	38
	Berlin surrounding with UHI imprint (2019, synthetic)	13,4	23,1	38,8	211	311	29	29	49

Fig. 4.6: Key Characteristics of Meteorological Data Used in Building Performance Simulations (BPS). This table presents selected meteorological parameters, color-coded from blue to red to indicate minimum to maximum values respectively (Berlin data not coloured). Definitions: Data covers the period from June 1 to August 31; Hot days are defined as days with a maximum outdoor air temperature > 30 °C; Tropical nights are days with a minimum outdoor air temperature ≥ 20 °C; Warm nights are days with a minimum outdoor air temperature ≥ 18 °C. * Summer: June 1st to August 31st

In this analysis, the primary question is whether different meteorological datasets significantly affect the assessment of overheating risk in the chosen application case of an attic bedroom, focusing on the room's heat resilience. To explore this, two distinct Overheating Analysis (OA) methodologies were implemented:

- *German Standard DIN 4108-2 Methodology*: This approach calculates the Temperature-Weighted Exceedance Hours (TWEH) above a set indoor temperature limit—27 °C for Dresden and 26 °C for Berlin—throughout the year. A positive OA outcome is determined if the TWEH does not exceed 1200 Kh/a.
- *Adaptive Criteria Approach*: This methodology explained in the last book chapter utilizes an adaptive temperature limit and differentiates between overall and nocturnal overheating. To achieve a favorable OA result, three specific criteria

must be met. The two criteria pertain to the permissible exceedance of indoor temperatures during the day, considering every day of the year, while a third criterion focuses on the allowable maximum nighttime temperature.

The following illustration Figure 4.7 presents a summary of the building simulation results using these two different evaluation criteria.

Category of comparison	Meteorological data set	OA accord. DIN 4108-2		OA according to proposed indicator set			
		Crtieria	OA result	Criteria I (whole day)	Criteria IIa (night)	Criteria IIb (night)	OA result
		TWEH (Kh/a)	-	TWEH (Kh/a)	TWEH (Kh/a)	ΔT (K)	-
1. Deviations of annual summer conditions	coolest summer for Dresden region (measured), year 1996	112	OA passed	82	3	1,0	OA passed
	average summer for Dresden region (measured), year 2007	699	OA passed	400	59	4,5	OA failed
	hottest summer for Dresden region (measured), year 2019	1882	OA failed	889	161	5,0	OA failed
2. Standardised meteorological input	former **TRY2010-12** for Dresden region accord. DIN4108-2	949	OA passed	503	85	3,6	OA failed
	former **TRY2010-04** for Dresden region accord. to DWD	126	OA passed	73	12	1,5	OA passed
	current **TRY 2017** for Dresden centre	973	OA passed	558	104	2,9	OA failed
	ASHRAE IWEC 2 for Dresden region	414	OA passed	209	35	2,3	OA failed
3. Considering climate change	**Year 2035** from former **TRY2010-04** for Dresden region	744	OA passed	163	13	1,1	OA passed
	Year 2045 from current **TRY2017** for Dresden centre	1182	OA passed	638	115	2,8	OA failed
	Year 2040 by our approach for Dresden region	1773	OA failed	824	193	3,5	OA failed
4. Considering urban heat island effect	Berlin surrounding - no UHI (2019, measured)	3027	OA failed	744	126	4,4	OA failed
	Berlin centre (2019, measured)	4516	OA failed	1253	271	4,9	OA failed
	Berlin surrounding with UHI imprint (2019, synthetic)	6269	OA failed	1934	486	5,8	OA failed

Fig. 4.7: Comparative Results of Overheating Analysis Using Different Meteorological Inputs. Green = Criteria fully met, Yellow: Criteria partially met, Red: Criteria failed.

The evaluation of overheating analysis (OA) methodologies applied to Dresden and Berlin using different meteorological datasets highlights contrasting outcomes and underscores the influence of dataset selection on simulation results. The German Standard DIN 4108-2 Methodology involves calculating Temperature-Weighted Exceedance Hours (TWEH) above a specified indoor temperature limit (27 °C for Dresden and 26 °C for Berlin) and generally shows favorable outcomes for Dresden. Most meteorological datasets, including those incorporating climate change and extreme summer conditions from 2019, indicated that Dresden met the overheating criteria. In contrast, the datasets from 2019 for Berlin invariably result in non-compliance with the established criteria. The Adaptive Criteria Approach, in stark contrast to the DIN 4108-2 method, a more nuanced approach often results in negative OA outcomes for Dresden across almost all meteorological conditions. This method, which applies adaptive temperature thresholds, frequently fails due to non-compliance with Criteria IIb—specifically, the requirement that indoor temperatures should not exceed the adaptive threshold by more than 2 °C each day during summer.

The differing results from these two methodologies illustrate the significant impact of choosing specific meteorological inputs on OA outcomes. Notably, selecting regional-specific data, such as TRY2010 for region 4 as recommended by the German Weather Service (DWD) for Dresden, leads to notable discrepancies in OA results. This emphasizes the necessity to carefully choose regional settings in OA methodologies to ensure accurate assessments. Moreover, incorporating factors like the Urban Heat Island (UHI) effect and future climate projections is crucial, as these elements substantially increase the potential for indoor overheating. Such considerations are vital for the long-term sustainability of urban buildings, highlighting the need for adaptive and forward-looking approaches in building performance simulations.

5 Conclusion

The evaluation of climate data's role in building overheating assessments is crucial for understanding and mitigating the risk of indoor overheating across Europe. This chapter highlights the dynamic relationship between climate conditions and the risk of overheating. Geographical factors such as latitude, proximity to the sea, and altitude significantly influence local temperature and radiation conditions, creating distinct microclimates that affect thermal comfort within buildings.

Urban climatic features, such as the Urban Heat Island (UHI) effect, further complicate this scenario. Urban areas, with their heat-retaining structures and reduced vegetation, experience higher temperatures compared to rural areas, exacerbating the risk of overheating. The interplay between anthropogenic heat sources, such as traffic and HVAC systems, and structural elements like dark roofs and pavements, contributes to elevated urban temperatures.

The *selection of appropriate climate datasets* is a critical challenge in building performance simulation (BPS). These datasets must represent long-term typical summer conditions for the building's location. Different datasets, even within the same location, can yield varying risk assessments due to differences in parameters like maximum temperature, solar radiation, and the frequency of hot days and tropical nights. The example of Dresden demonstrated the multitude of metrics that can form the basis for selection. However, the impact of climate datasets on a building is also strongly influenced by the specific characteristics of the building itself (such as the orientation and size of window areas) and its usage. Therefore, reliable statements cannot be made based solely on individual weather elements or exemplary simulation calculations.

However, verified measurements at specific building sites are often unavailable, and meteorological conditions can vary widely. Common practices include using data from meteorological stations or standardized synthetic datasets, such as Test Reference Years (TRY) for Germany, Design Summer Years (DSY) for the UK, or IWEC 2 data for international contexts. The *study in Section 4* integrated a diverse array of meteorological datasets in Building Performance Simulation to *evaluate the differences in overheating assessment* of an attic bedroom in Dresden and Berlin, Germany. The datasets encompass measured data from meteorological stations, standardized meteorological data, and data accounting for climate projections and the local UHI effect. Our findings unequivocally demonstrate that the overheating risk in buildings is profoundly contingent on the choice of meteorological data sets. While the natural annual fluctuations in measured summer conditions from 1991 to 2020 (ranging from cool to hot summers) induce notable disparities in overheating intensity, the incorporation of localized climate conditions and climate projections is deemed essential for an accurate overheating risk assessment. This underscores the *significance of developing and refining locally resolved datasets*, which, due to the limitations of measuring stations, must not only **rely on meteorological measurements** but also **incorporate urban climate simulations and, for future scenarios,** suitable downscaling methods of climate model outputs. Although initial strides in this direction have been successful, further research is imperative to generate more reliable meteorological inputs for robust OA, tailored to future summers, and considering both regional effects and local UHI impacts for larger cities.

Effective **mitigation of overheating risks** requires **careful selection and consideration of local climatic nuances** in building design and urban planning. The use of high-quality, representative climate datasets is essential for accurate simulation and assessment. Additionally, incorporating future climate projections ensures that buildings are resilient under changing climate conditions. Urban planning should focus on increasing vegetation, optimizing building materials, and enhancing natural ventilation to mitigate the UHI effect and improve thermal comfort.

6 References

1. **European Environment Agency (ECA).** *Indices of extreme temperature indicators.* [Internet]. [cited 2024 Feb 16]. Available from: https://www.eea.europa.eu/data-and-maps/data/external/indices-of-extremes-temperature-indicators
2. **National Centers for Environmental Information (NCEI).** *Global Hourly - Integrated Surface Database (ISD).* [Internet]. 2021 [cited 2024 Apr 25]. Available from: https://www.ncei.noaa.gov/products/land-based-station/integrated-surface-database
3. **Rousi E, Kornhuber K, Beobide-Arsuaga G, Luo F, Coumou D.** *Accelerated western European heatwave trends linked to more-persistent double jets over Eurasia.* Nat Commun [Internet]. 2022 Jul 4 [cited 2024 Feb 16];13(1):3851. Available from: https://www.nature.com/articles/s41467-022-31432-y
4. **European Space Agency.** *Europe braces for sweltering July.* [Internet]. [cited 2024 Mar 7]. Available from: https://www.esa.int/Applications/Observing_the_Earth/Copernicus/Sentinel-3/Europe_braces_for_sweltering_July
5. **Eurostat.** *Half of EU's population older than 44.4 years in 2022.* [Internet]. 2023 Feb [cited 2024 Mar 7]. Available from: https://ec.europa.eu/eurostat/web/products-eurostat-news/-/ddn-20230222-1
6. **Eurostat.** *Median age ranges from 18 to 54 across EU regions.* [Internet]. [cited 2024 Mar 7]. Available from: https://ec.europa.eu/eurostat/web/products-eurostat-news/-/ddn-20170215-1
7. **Ballester J, Quijal-Zamorano M, Méndez Turrubiates RF, Pegenaute F, Herrmann FR, Robine JM, et al.** *Heat-related mortality in Europe during the summer of 2022.* Nat Med [Internet]. 2023 Jul [cited 2023 Aug 24];29(7):1857–66. Available from: https://www.nature.com/articles/s41591-023-02419-z
8. **Tandon A.** *Heat-related deaths 56% higher among women during record-breaking 2022 European summer.* [Internet]. 2023 Jul [cited 2024 Mar 7]. Available from: https://www.preventionweb.net/news/heat-related-deaths-56-higher-among-women-during-record-breaking-2022-european-summer
9. **De Pauw K, Depauw L, Calders K, Caluwaerts S, Cousins SAO, De Lombaerde E, et al.** *Urban forest microclimates across temperate Europe are shaped by deep edge effects and forest structure.* Agricultural and Forest Meteorology [Internet]. 2023 Oct 15 [cited 2024 Mar 9];341:109632. Available from: https://www.sciencedirect.com/science/article/pii/S0168192323003234
10. **Kuttler W, Weber S.** *Characteristics and phenomena of the urban climate.* Meteorologische Zeitschrift [Internet]. 2023 Jun 26 [cited 2024 Jun 18];15–47. Available from: https://www.schweizerbart.de/papers/metz/detail/32/102631/Characteristics_and_phenomena_of_the_urban_climate
11. **Mentaschi L, Duveiller G, Zulian G, Corbane C, Pesaresi M, Maes J, et al.** *Global long-term mapping of surface temperature shows intensified intra-city urban heat island extremes.* Global Environmental Change [Internet]. 2022 Jan 1 [cited 2024 Mar 1];72:102441. Available from: https://www.sciencedirect.com/science/article/pii/S095937802100220X
12. **Wetter und Klima - Deutscher Wetterdienst.** *Leistungen - Open Data.* [Internet]. [cited 2024 Apr 26]. Available from: https://www.dwd.de/DE/leistungen/opendata/opendata.html
13. **Larsen TS, Heiselberg P.** *Single-sided natural ventilation driven by wind pressure and temperature difference.* Energy and Buildings [Internet]. 2008 Jan 1 [cited 2024 Feb 29];40(6):1031–40. Available from: https://www.sciencedirect.com/science/article/pii/S0378778807002137

14. **Oke TR, Mills G, Christen A, Voogt JA.** *Urban Climates.* Cambridge: Cambridge University Press; 2017 [cited 2024 Apr 26]. Available from: https://www.cambridge.org/core/books/urban-climates/A02424592E1C7F9B9CD69DAD57A5B50B

15. **Deutscher Wetterdienst.** promet, meteorologische Fortbildung, Heft 106: *Stadtklima im Wandel.* 2023 [cited 2024 Apr 26]; Available from: https://www.dwd.de/DE/leistungen/pbfb_verlag_promet/l_promethefte/106p.html

16. **Maraun D, Shepherd TG, Widmann M, Zappa G, Walton D, Gutiérrez JM, et al.** *Towards process-informed bias correction of climate change simulations.* Nature Clim Change [Internet]. 2017 Nov [cited 2024 Mar 12];7(11):764–73. Available from: https://www.nature.com/articles/nclimate3418

17. **Knutti R, Sedláček J, Sanderson B, Lorenz R, Fischer E, Eyring V.** *A climate model projection weighting scheme accounting for performance and interdependence.* Geophysical Research Letters. 2017 Feb 1;44. Available from: https://doi.org/10.1002/2016GL072012

18. **Deutsches Institut für Normung (DIN).** DIN 4108-2:2013-02, *Wärmeschutz und Energie-Einsparung in Gebäuden - Teil 2: Mindestanforderungen an den Wärmeschutz.* [Internet]. Available from: https://www.beuth.de/de/-/-/167922321

19. **ASHRAE (American Society of Heating, Refrigerating and Air-Conditioning Engineers).** *ASHRAE 55:2023 Thermal Environmental Conditions for Human Occupancy.* [Internet]. ASHRAE 55:2023 2023. Available from: https://www.ashrae.org/technical-resources/bookstore/standard-55-thermal-environmental-conditions-for-human-occupancy

20. **Krähenmann D.** *Handbuch Ortsgenaue Testreferenzjahre von Deutschland für mittlere, extreme und zukünftige Witterungsverhältnisse.* Offenbach: Deutscher Wetterdienst und Bundesinstitut für Bau- Stadt- und Raumforschung; 2017 Jul p. Available from: https://www.dwd.de/DE/leistungen/testreferenzjahre/testreferenzjahre.html

21. **EQUA.** *IDA ICE - Simulation Software.* [Internet]. 2023 [cited 2024 Apr 26]. Available from: https://www.equa.se/de/ida-ice

22. **Kriesten T, Ziemann A, Goldberg V, Schünemann C.** *The Effect of Regional, Urban and Future Climate on Indoor Overheating – a Simplified Approach Based on Measured Weather Data, Statistical Evaluation, and Urban Climate Effects for Building Performance Simulations.* 2024. Available from: https://papers.ssrn.com/sol3/papers.cfm?abstract_id=4798151

23. **Wienert U, Kreienkamp F, Spekat A, Enke W.** *A simple method to estimate the urban heat island intensity in data sets used for the simulation of the thermal behaviour of buildings.* Meteorologische Zeitschrift [Internet]. 2013 Apr 1 [cited 2024 Apr 26];179–85. Available from: https://www.schweizerbart.de/papers/metz/detail/22/80287/A_simple_method_to_estimate_the_urban_heat_island_intensity_in_data_sets_used_for_the_simulation_of_the_thermal_behaviour_of_buildings

Peggy Freudenberg
Thermal Bearability and Thermal Comfort

Approaches for the Assessment of Indoor Climate Performance under Hot Conditions

Abstract: This chapter delves into the complex subject of defining tolerance and comfort thresholds for indoor climates, with a particular focus on the challenges of establishing these thresholds during summer. It proposes a methodology grounded in climate chamber studies to set explicit bearability threshold values tailored for healthy adults. The analysis extends to the heightened heat sensitivity of demographic groups such as the elderly, individuals with pre-existing health conditions, and young children, suggesting that their threshold values be derived from excess mortality data collected during European heatwaves—an area previously underexplored in indoor climate research. Additionally, the chapter reviews the current landscape of comfort models, assessing their suitability for summer comfort evaluation. It highlights the inadequacies of the Fanger-PMV model in accurately determining optimal summer indoor air temperatures in both air-conditioned and non-air-conditioned environments. The predominance of research focusing on office buildings in Europe exposes a gap in the applicability of these findings to other building types, such as residential homes, indicating the need for broader investigatory efforts. Moreover, the limitations of empirical and adaptive models are critically examined, noting their frequent failure to account for usage nuances, such as nighttime comfort. The chapter argues that these models often fail to fully capture the diverse and dynamic ways buildings are utilized and the varied thermal comfort needs of their occupants. The methodology employed in this chapter consists of a comprehensive literature review of empirical studies concerning tolerance thresholds and comfort models. This literature review is supplemented by the analysis of published datasets, including climate chamber studies on bearability and comfort databases. Through this dual approach, the research aims to provide a robust foundation for understanding and establishing indoor climate thresholds, especially in the context of summer heat in Europe.

Keywords: Thermal Comfort, Heat Sensitivity, Indoor Climate, Adaptive Thermal Comfort Models, Summer Comfort Assessment, Heatwave Vulnerability

Peggy Freudenberg, Dresden University of Technology, Institute of Building Climatology, Zellescher Weg 17, 01062 Dresden, +49(0)351 463-35259, peggy.freudenberg@tu-dresden.de

7   Introduction

The assessment of summer indoor climates poses a significant challenge due to the complex interplay of climatic conditions, building characteristics, and human physiology. This complexity is magnified when considering the diverse sensitivities and adaptive capacities of different demographic groups. This chapter aims to address critical questions surrounding the establishment of bearability and comfort thresholds for summer indoor climates and the specific needs of heat-sensitive populations. A significant contribution of this chapter is the clear delineation between thermal comfort and bearability. While bearability addresses the fundamental health safety requirements and should be mandated by regulations, thermal comfort focuses on the optimal conditions for occupant satisfaction, guiding the design and planning of indoor environments.

Bearability, in this context, refers to the *minimal requirements* for indoor climates that *ensure health safety*, even with prolonged exposure. Bearability should be addressed and ensured by higher authorities, such as national laws and regulations. Establishing these thresholds involves identifying appropriate values for the tolerability of summertime indoor climates. Another vital aspect of this chapter involves understanding which demographic groups are particularly vulnerable to heat and how indoor environments can be adapted to meet their needs. Heat sensitivity varies significantly among individuals, influenced by factors such as age, health status, and acclimatization to climate. Elderly individuals, young children, people with pre-existing health conditions, and those taking certain medications are typically more prone to heat stress. The formulation of specific threshold values for heat-sensitive groups requires a detailed analysis of heat wave mortality studies. By leveraging data from past extreme heat events and examining the physiological responses of these groups, the research intends to establish a set of actionable temperature limits. These limits should serve as a critical reference for building regulations and standards aimed at protecting the most vulnerable occupants from heat-related health risk.

Thermal comfort is defined according to the ASHRAE Standard 55, which describes it as 'that condition of mind which expresses *satisfaction with the thermal environment*' (1,p.55). Thermal comfort should be a key consideration in the planning and design phases of building projects, often determined through building simulation calculations. This chapter explores the extent to which established comfort models, such as the Fanger Predicted Mean Vote (PMV) model, and empirical (adaptive) models are suited for application under specific climatic conditions, particularly during the summer, and across different building types. It will investigate whether these models adequately reflect the thermal comfort requirements in varied environments and identify potential areas where they may fall short. The suitability of these models for different settings, such as air-conditioned versus naturally ventilated buildings, and their effectiveness in capturing the diverse and dynamic usage patterns of buildings will be critically examined.

Through extensive literature analysis and insights into underlying datasets, this chapter aims to advance our understanding of thermal comfort in summer, propose practical approaches for indoor climate assessment, and contribute to the health and well-being of building occupants across diverse environments.

8 Thermal Bearability

Thermal bearability of indoor climates is fundamentally about **setting the minimum standards** necessary to ensure that **environments are usable** without posing health risks to their intended occupants. While such standards partially exist for winter conditions and residential buildings in countries like Germany, Poland, France, the UK, and Sweden, comparable guidelines or regulations for summer conditions are largely absent. An exception is the German ASR, which legally mandates a maximum indoor air temperature of 35 °C for workplaces (2).

The other European regulations are non-binding recommendations focusing on the comfort level. The recommended values range between 25 and 28 °C, with different approaches to permissible exceedance. In some cases, the user group is taken into account in the selection of limit values (Sweden: differentiation between older and younger users) and in others, the type of room is taken into account (UK: e.g. differentiation between living and sleeping areas)(3–5).

While establishing **summer indoor temperature limits** has clear benefits, it also presents significant challenges and limitations. One critical issue is the high dependence on unpredictable factors attributed to the building occupancy, highlighted in (6). Unlike winter conditions, which are more uniform and predictable, summer temperatures can vary widely based on geographic location, building orientation, and occupancy patterns. Additionally, unlike winter heating systems, many European buildings are typically not equipped with air conditioning systems capable of maintaining specific summer temperature limits (7). This lack of infrastructure means that implementing strict summer thresholds may not be feasible. Therefore, the standards for summer cannot be as restrictive as those for winter. A flexible approach to setting maximum temperatures is included in the British CIBSE TM52 (4), where the limit of the indoor operative temperature is set at a difference of 4 K above the moving average of the outdoor air temperature, as referenced in EN 15251. Although this approach is not directly formulated as a bearability criterion, it can be interpreted as such.

However, it is crucial to develop a strategy that enhances the robustness and resilience of buildings in response to climatic trends outlined in Chapter 1. Historically, the severe impact of extreme winters, such as the 'Jahrhundertwinter' of 1929 in Germany, which resulted in thousands of deaths due to inadequate heating and insulation, gradually led to the implementation of stringent building standards (8). Initially, responses were piecemeal and varied, but over time, the necessity of robust thermal protection became evident, culminating in comprehensive regulations to ensure minimum indoor temperatures during winter months. Similarly, recent decades have seen severe heatwaves causing significant mortality, yet the response in terms of regulatory updates for summer conditions has been comparatively limited.

A pertinent method to explore the bearability of indoor climates involves examining the relationship between **temperature levels** and **heat-related mortality** during heatwaves. Historically, Europe has experienced heatwaves of differing lengths and intensities since 1950 depending on the region, with particularly severe events occurring in Russia in 2010, Central Europe in 2003, Finland in 1972, and the UK in 1976 (9). Subsequent notable heatwaves in Central Europe appeared in 2006 and 2015. The impact of these events has varied significantly across affected nations, as illustrated in the following summary in Figure 8.1.

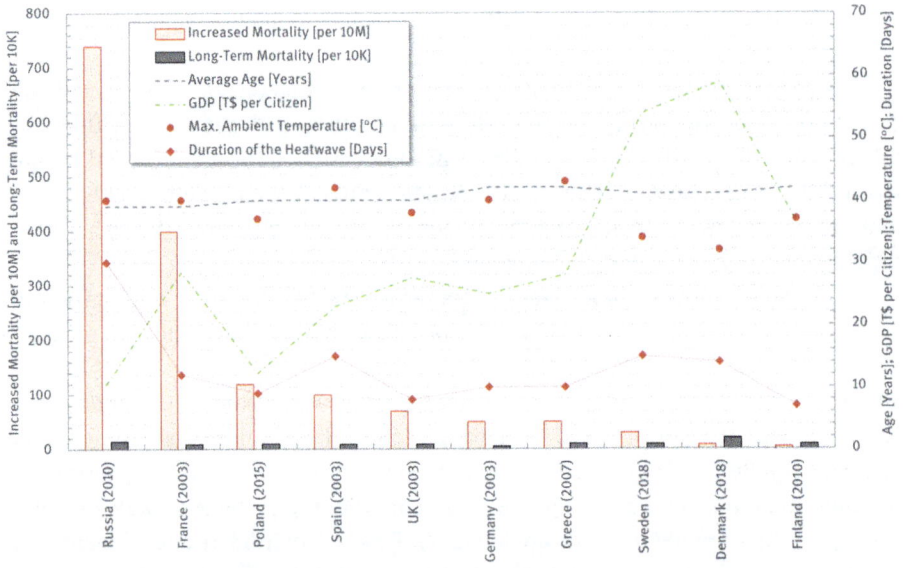

Fig. 8.1 Overview of Significant European Heatwaves Since 2000, Based on Data in (9,10).

Figure 8.1 highlights the dramatic increase in mortality during the severe 2010 Russian heatwave, where air temperatures frequently exceeded 40 °C over an extended period. This heatwave also impacted several other countries, including Ukraine, Mongolia, and China. The heightened mortality rates can be attributed not only to the extreme temperatures but also to widespread forest and peat fires, compounded by persistent atmospheric inversions. These inversions severely degraded air quality in urban areas and near fire zones, leading to particularly catastrophic conditions in large cities like Moscow (9,11,12). These significant heatwaves all **lasted more than a week**, with maximum **temperatures ranging from approximately 32 °C to 44 °C** throughout the duration of each event.

Although the data for heatwave-related excess mortality are based on measured outdoor air temperatures, which are not directly equivalent to indoor air temperatures, it can be inferred that the thermal storage capacity of buildings would be exhausted after several days of sustained high temperatures. Consequently, even with an adequate structural overheating protection, indoor air temperatures would at least reach the same levels as the recorded outdoor temperatures.

8.1 Threshold Values Derived from Mortalities During Heatwaves

When examining the correlation between heat-related excess mortality and variables such as the average age, prosperity, and climatic conditions of different regions, it becomes apparent that no single factor effectively predicts the impact of heatwaves on populations. However, the analyses of these events reveal that the **duration of the heat phase is a pivotal risk factor**. This is evident in Figure 2.1 which shows that the exceptionally long heat phases, ranging from 8 to 30 days, correlate with higher excess mortality. In contrast, the European heatwaves of 2006, 2010, and 2015, which were much shorter, resulted in significantly lower excess mortality (9,13). Other **significant factors** contributing to heat-related excess mortality include **high nocturnal temperatures** (above 20 °C), **high humidity levels, rapid temperature rises**, and **preceding cold spells** (12–14). Individual risk factors, such as age and pre-existing health conditions, are crucial in determining susceptibility.

Numerous studies have documented increased mortality rates among older adults, individuals with chronic health conditions, very young children and infants, pregnant women, and people who are overweight. Similarly, socially disadvantaged individuals and certain occupational groups, particularly those working outdoors, are also at heightened risk. However, there is some debate regarding the influence of gender on mortality rates during heatwaves, with some research indicating higher rates for women (15,16), while other studies find no significant gender-based differences (17,18). **Maximum daily temperatures** are also linked to increased mortality. The threshold temperatures that are considered critical vary significantly by region and largely **depend on the typical summer temperatures experienced locally**. For example, in London, excess mortality increased significantly when daily maximum temperatures exceeded **27 °C** . In contrast, in Valencia, the critical threshold was around 40 °C. For Central Europe, temperatures exceeding **31–36 °C** were deemed critical (10,12–14).

This data supports the hypothesis that populations in areas where high temperatures are more common are likely better adapted to these conditions, whether through physiological acclimatization or adjustments in behaviour and infrastructure. Such infrastructural adaptations might include urban green spaces and parks, densely built-up inner-city areas with sheltered pedestrian zones, buildings designed to resist heat with thermal mass and reduced window areas, and advanced cooling technologies. These measures tend to be more developed in regions that typically experience hotter summers and are prevalent in historical urban districts, as opposed to cooler regions or in more modern constructions. In conclusion, the combination of extended heat duration and specific regional conditions significantly affects heatwave mortality rates.

8.2 Threshold Values Derived from Human Thermoregulation

Determining the tolerability limits for indoor climates can also be approached through the lens of human physiology and thermoregulation. Research in this area is divided into studies focusing on outdoor heat stress and those concerning indoor environments. For outdoor conditions, one of the key parameters used to assess heat stress is the Wet Bulb Globe Temperature (WBGT) index (19,20). This index considers factors including radiant heat exposure, which is a significant component of heat stress in outdoor settings but is less pertinent indoors. Indoors, simpler metrics are used such as the ***Wet Bulb Temperature (WBT)***, which evaluates the combined effects of air temperature and humidity.

A critical limitation of these simplified indoor models is their tendency to overlook or oversimplify individual factors that significantly impact heat stress. These factors include metabolic heat production due to physical activity and the insulating effect of clothing, which can hinder heat dissipation. To address these shortcomings, two strategies can be employed. First, setting simplified tolerability limits specific to different demographic groups can help account for variability in susceptibility to heat stress. Secondly, establishing tolerance values based on a heat balance model of the human body offers a more holistic approach. This approach calculates the net heat exchange between the body and its environment, allowing for a nuanced understanding of how different conditions affect thermal comfort and stress. An advanced method to evaluate heat stress in indoor environments is the ***Predicted Heat Strain (PHS)*** model, as defined in the ISO 7933 standard (21,22). The PHS model integrates various physiological and environmental factors to predict the potential for heat strain and related health risks in working environments. By calculating the predicted core temperature and sweat rate, it offers a comprehensive assessment of thermal stress. However, despite its comprehensive nature, the PHS model faces several criticisms. It requires detailed and precise input data, making it difficult to apply in real-world settings without sophisticated equipment and expertise. Additionally, the PHS model often over-predicts the risk of heat strain, leading to overly cautious measures that may not reflect actual conditions accurately (23–25).

8.2.1 Threshold Values for Young Healthy Adults

To derive the **limits for bearable conditions**, an abstract and simplified approach using the Wet Bulb Temperature (WBT) will be employed. This method provides a fundamental understanding of how temperature and humidity interact to affect human comfort and safety, without the complexity of more detailed models. The derivation of a WBT limit value fundamentally relies on the principle that the human body must maintain its core and skin temperatures at roughly 37 °C and 35 °C, respectively. This temperature difference is crucial for the effective dissipation of heat from the body's core (19,26). In environments where the air is not saturated with humidity, the body can dissipate heat through sweating and the subsequent evaporation from the skin surface. This evaporative cooling mechanism is essential as it significantly enhances the body's ability to regulate its core and skin temperatures, even when ambient temperatures rise above 35 °C, where radiant and convective heat emission become ineffective. This process depends heavily on the air's ability to absorb moisture; drier air facilitates better heat loss from the body. However, in conditions of saturated air, where the air holds as much moisture as it can, the body's ability to lose heat through evaporation is severely compromised. Given these limitations in heat dissipation mechanisms, and assuming the body's heat storage capacity is not a factor (i.e., under long-term or steady-state conditions), it becomes essential that the surrounding air temperature must remain **below 35 °C in moisture-saturated environments** to prevent the body's core temperature from exceeding safe levels. As the relative humidity decreases, evaporative cooling through perspiration becomes more effective, allowing for higher air temperature tolerances. The WBT is intended to represent the body's capacity for heat dissipation to the environment; the closer it approaches 35 °C, the more limited the body's ability to release heat.

However, when this theoretical framework is juxtaposed with actual heat stress occurrences, its applicability appears limited (27,28). Studies in (28) indicate that for young adults, the bearability limits are breached at a wet bulb temperature of 30–31 °C in dry conditions and just 25–28 °C in humid conditions, at which points the body struggles to maintain its core and skin temperatures during moderate physical activity. This disparity suggests that the **WBT alone, without additional humidity information, may not be a reliable indicator of heat stress**.

The insights gleaned from this empirical research have been integrated into a threshold function for environmental bearability based on air temperature and relative humidity as shown in Figure 8.2. This threshold function serves as a pivotal tool for assessing the interaction between temperature and humidity in determining the limits of human comfort and safety in a simplified manner.

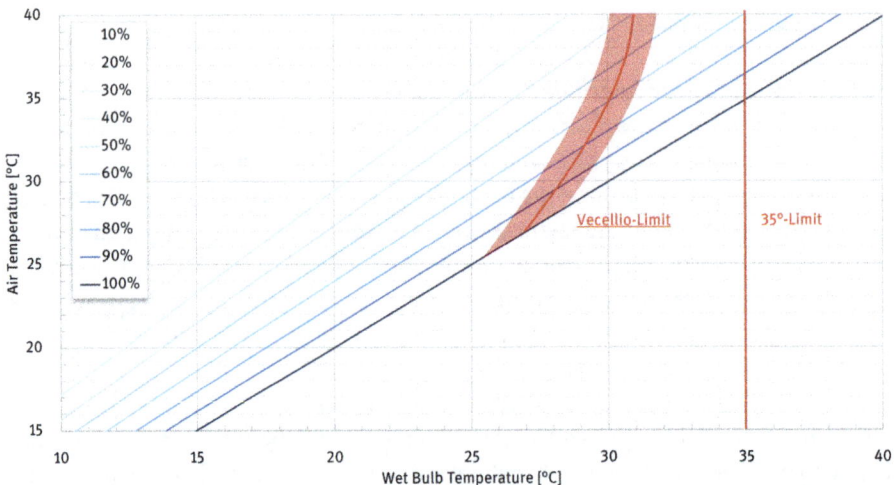

Fig. 8.2: Graphical Representation of Wet Bulb Temperature as a Function of Air Temperature and Relative Humidity, Based on Vecellio's Experimental Results in (28) (Y-axis: Air Temperature; Blue Curves: Relative Humidity from 10% to 100%).

Translating the findings from this study to the climatic conditions of Central Europe, where summer humidity typically ranges between 60–80%, a specific range of tolerable air temperatures can be established for healthy adults engaged in moderate physical activity while wearing light clothing. Given these conditions, the ideal air temperature range for maintaining comfort and safety is **approximately 32 °C to 37 °C**, with an airspeed of about 0.5 meters per second. Considering the variability observed in the study results, this temperature range could reasonably be extended from about 31 °C to 38 °C to accommodate different individual tolerances and specific environmental variations. This adjusted range helps ensure that thermal comfort guidelines are both practical and inclusive of a broader spectrum of scenarios encountered during typical Central European summers.

8.2.2 Threshold Values for Vulnerable People based on Mortalities

Research into threshold values for thermal environments has also encompassed demographic groups other than healthy, young individuals. However, many of these studies have focused more on understanding the unique thermoregulatory characteristics of these groups rather than establishing precise environmental temperature limits. This focus aligns with a broader trend in empirical research on heatwave-related excess mortality, where the primary interest often lies in exploring the relative vulnerabilities among different groups rather than defining specific threshold values.

In these investigations of excess mortality, secondary analyses of mortality outcomes primarily derived from hospital statistics were conducted rather than primary data analysis. Consequently, individual parameters crucial for heat balance, such as activity levels and clothing, cannot be reconstructed. Similarly, specific details regarding pre-mortem locations or climate conditions are not available. The physiological and empirical insights into these particular groups are summarized as follows:

- **Older People:** Older adults often face challenges such as diminished cardiovascular capacity and reduced sweat production, impacting their thermoregulation. Compounding factors like decreased thirst perception and the effects of certain medications can lead to inadequate hydration (18–21). Evidence suggests that heat-related excess mortality in individuals aged 50 and older begins to increase significantly at air temperatures around 29–30 °C (15,29–32). Therefore, it is advisable to keep maximum air temperatures below this range for the elderly.
- **People with Chronic Conditions:** Individuals with pre-existing cardiovascular and respiratory diseases experience additional strain during thermal regulation due to the circulatory demands. For this group, maintaining maximum air temperatures below 30 °C is recommended to mitigate health risks (33–36).
- **Children:** Infants and young children, particularly those under one year, have a compact body shape with a relatively small surface area for heat regulation. Their higher core temperature and underdeveloped thermoregulatory systems, combined with a higher metabolic rate, make them more susceptible to heat-related issues. It is suggested that the maximum air temperature for this age group should not exceed 29 °C (13,17,37).
- **Severely Overweight Individuals:** Higher body mass index (BMI) can impair thermoregulation due to a proportionally smaller body surface area and the insulating properties of body fat, which impedes heat dissipation and increases heat storage, particularly during physical exertion. While no specific temperature guidelines have been established for this group, the challenges they face highlight the need for cautious environmental management (13,13,32).
- **Pregnant Women:** Pregnancy increases heart rate and blood volume, placing additional stress on the cardiovascular system. High body temperatures have been linked to complications such as fetal development issues, premature births, miscarriages, and stillbirths, particularly in the first trimester. Research indicates an increase in miscarriage rates at short-term maximum air temperatures of 32°C, with risks escalating as temperatures rise. Consequently, it is prudent to maintain air temperatures below 32 °C for pregnant women (38–41).

The impact of heatwaves extends beyond these specific groups, affecting anyone who is physically active or spends significant time outdoors. Addressing these risks involves not only structural measures for heat protection but also educational and occupational health and safety initiatives to reduce exposure and enhance resilience. Figure 8.3 provides a comprehensive visualization of the threshold values for various demographic groups, summarizing the findings discussed in the previous sections. For each group—older adults, individuals with chronic conditions, infants, overweight individuals, and pregnant women—the figure displays the temperatures that correspond to increased heat-related health risk.

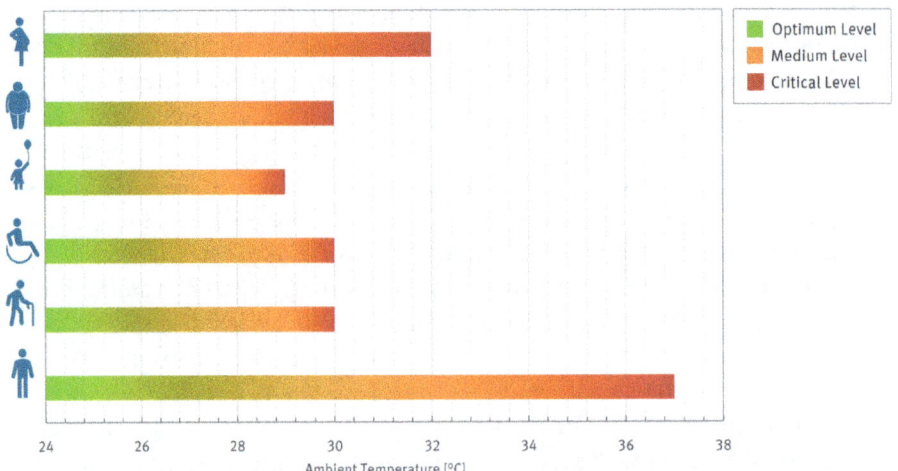

Fig. 8.3: Thermal Bearability Scales (from top to bottom): Pregnant Women, Obese Individuals (estimated, precise value ranges not specified in current literature), Children, People with Diseases and Disabilities, Elderly, and Young Healthy Adults.

In summary, the analysis reveals that while young, healthy adults in Central Europe can tolerate temperatures of 32–37 °C, more **vulnerable groups require lower thresholds of 29–32 °C** due to varying heat sensitivities. Older individuals and people with pre-existing health conditions, particularly those with cardiovascular and respiratory issues, are at risk at temperatures above 29–30 °C. The same limit is assumed for obese individuals. Infants require temperatures below 29 °C for safe thermoregulation, while pregnant women should be kept below 32 °C.

It is important to recognize that the maximum outdoor air temperatures identified in these studies are not directly equivalent to indoor air temperatures. However, during prolonged heatwaves exceeding a week, it is reasonable to assume that indoor climatic conditions will at least mirror those outdoors, particularly in the absence of active cooling systems or adaptive architectural features. This assumption is critical for understanding the impact of external heat on indoor environments. In scenarios where buildings lack heat-adaptive architecture, indoor temperatures may actually exceed those recorded outdoors. This discrepancy highlights a simplification inherent in this approach, as it does not fully account for the potential exacerbation of indoor heat conditions in poorly adapted structures.

9 Thermal Comfort

Bearability represents the minimum standards required to avoid adverse health effects, setting a baseline that ensures health safety rather than optimal well-being. In contrast, comfort describes the *ideal indoor climate*, focusing on optimizing conditions to enhance occupants' quality of life and productivity. According to both the European standard EN 16798-1:2019 and the ASHRAE Standard 55:2020, comfort is defined as '[...] condition of mind which expresses satisfaction with the thermal environment and is assessed by subjective evaluation.' (1,42). These standards emphasize the importance of creating indoor environments that balance thermal conditions to achieve optimal comfort and overall well-being.

Ensuring a comfortable indoor climate is not merely a modern-day architectural challenge but a fundamental aspect that has been integral to building design for centuries. The concept of comfort in relation to ambient climate has deep historical roots, notably explored by Alexander von Humboldt. Humboldt's holistic definition of 'climate' included a range of atmospheric changes perceivable by human senses. He detailed factors such as temperature, relative humidity, air pressure, air speed, and solar radiation, which he documented in his comprehensive studies (43). These elements, identified by Humboldt, continue to form the core considerations in the evaluation of thermal comfort within indoor environments. Humboldt particularly underscored the *critical role of temperature* in understanding climatic impacts. Building on his foundational work, later methodologies expanded the scope to include air temperature along with *additional environmental factors*. Notable among these are various heat indices developed to assess thermal environments more accurately, such as the Effective Temperature (ET*) approach (44) and the Heat Stress Index (HSI) (45). These indices incorporate the wet bulb temperature, which has been proven effective in signalling heat stress conditions exemplified by a specific characteristic value (20,46,47).

However, as research evolved, it became apparent that simply aggregating ambient climate parameters was insufficient for accurately assessing indoor comfort. The initial models were expanded to *include individual characteristics such as clothing and metabolic activity*, yet these too failed to capture the full complexity of human comfort perception. This realization highlighted several significant limitations of the early models. Researchers noted the tendency of these models to overestimate certain indoor climate elements, a lack of consideration for temporal dynamics in environmental conditions, and challenges in applying these findings across diverse populations (19).

9.1 Heat Balance-Based Thermal Comfort Models

The advancements in thermal comfort research by **Povl Ole Fanger** marked a significant shift in understanding indoor climate control (48,49). Fanger introduced an innovative method that not only considered the most-relevant indoor climate conditions but also integrated the thermoregulatory responses of the human body. His
approach involved a structured, though simplified, heat balance model of the body,
which was a critical improvement over previous method. Fanger's second major contribution was his methodological use of climate chamber studies to align the outcomes of the heat balance model with actual human perceptions of thermal comfort.
These extensive studies allowed for a more accurate and empirically validated understanding of how people experience different thermal environments. The culmination
of Fanger's research was the development of the **Predicted Mean Vote** (PMV) model.
This model uses six key variables: air temperature, radiant temperature, air velocity,
relative humidity, the insulation value of clothing, and metabolic rate. These inputs
are used to calculate a thermal sensation parameter that ranges on a scale from -3
(indicating extreme cold) to +3 (indicating extreme heat), with 0 representing optimal
thermal equilibrium. The PMV model has become a foundational tool in the field of
environmental ergonomics, offering a quantifiable measure of human thermal comfort that can be applied in various settings to enhance indoor environmental quality.
Povl Ole Fanger's approach was revolutionary in the field of thermal comfort and indoor climate studies. His model integrated and synthesized numerous findings from
other researchers concerning human heat balance and related input parameters. His
heat balance-based model quickly gained popularity and was soon incorporated into
standardization processes. Remarkably, this assessment approach has been an integral part of ISO 7730 (50) since the early 1980s (51) and has remained unchanged since
then. Fanger's model also forms a critical part of the US ASHRAE 55 standards, underscoring its wide acceptance and enduring relevance in designing comfortable indoor environments.

Systematic **reviews of the PMV model**, particularly in air-conditioned environments within hot climates, exposed **significant limitations** in its application (52,53).
According to Avelino et. al. (52), only about 20% of the studies considered the PMV
model effective in accurately predicting thermal sensation votes (TSV) in such settings. A key issue identified is the PMV model's **tendency to overestimate heat sensation** and underestimate cold on its 7-point scale, which leads to inaccuracies in
these specific climatic conditions. The discrepancies in the PMV model's effectiveness are largely attributed to inaccuracies in estimating metabolic rates and the failure to incorporate personal factors like gender. The same conclusion was drawn by
Cheung et. al. (53) in the evaluation of the PMV model based on data from the ASHRAE
Global Comfort Database. Furthermore, the **rigidity of the PMV** model in predicting
comfortable indoor temperatures often results in recommendations that are not only
below the physiological or subjective comfort thresholds but also lead to increased

energy consumption. This ***over-cooling during the summer*** months is economically and environmentally costly, as it prompts a higher than necessary energy use.

The findings suggest, on the one hand, that the PMV model fails to account for the adaptation of populations in hot, humid regions who have developed a higher tolerance for heat. On the other hand, the ***thermal sensation scale*** (TSV), which relies on subjective self-reporting, can lead to ***underestimation of thermal discomfort***, particularly at higher temperatures, as shown by Schweiker et al. (54). Individuals may adapt to warmer conditions and report lower levels of discomfort than what objective measures would suggest. While the overestimation by the Fanger PMV model and the underestimation by the TSV scale may somewhat balance each other out, this balancing effect can vary significantly based on situational factors. This circumstance necessitates a closer examination of the application of both the model and the scale.

9.2 Empiric Thermal Comfort Models

In settings where ***buildings*** are ***not centrally air-conditioned***, such as those allowing occupants the freedom to modify their environment through actions like opening windows, utilizing sunshades, adjusting their attire, or using fans, the heat balance-based ***Fanger model*** tends to be ***even less effective***. This is primarily due to two critical insights gained from empirical observations. Firstly, it has been recognized that individuals are not merely passive recipients of their surrounding climate conditions. Instead, they actively engage in modifying their immediate environment to enhance their thermal comfort. This can include adjustments in clothing, altering their physical activity, or directly manipulating their immediate surroundings like airflow or sunlight exposure. Secondly, the tolerance of individuals to deviations from what might be considered the optimal indoor climate settings is significantly broader than what heat balance-based models typically predict. This ***greater tolerance*** is not just due to physiological capabilities but is also influenced by psychological factors, such as seasonal expectations, which heavily impact how indoor climate conditions are perceived (19,55–58). These observations highlight the limitations of traditional models that do not account for human behaviour and the dynamic interaction between people and their environments. This ***necessitates*** the development of ***more flexible and adaptive thermal comfort models*** that better reflect real-world scenarios of building use and human interaction with the built environment.

Adaptive model approaches derived from empirical studies in free-running buildings offer this nuanced understanding of thermal comfort, incorporating the concepts of behavioural, psychological, and physiological adaptation factors. These models have the potential to integrate the various ways individuals interact with their environments to maintain thermal comfort:

– ***Behavioral Adaptations***: Individuals engage in both conscious and subconscious actions to achieve thermal equilibrium. For instance, one might consciously put on a light jacket when the temperature drops or subconsciously move to a sunnier spot or adjust their posture to stay warm (58–60).

– **Psychological Factors:** These models consider how an individual's background and previous thermal experiences influence their comfort perceptions. For example, someone from a colder climate may tolerate lower indoor temperatures better than someone from a warmer climate due to their acclimatisation and expectations (58–60).

– **Physiological Responses:** This includes long-term changes in physiological responses due to prolonged exposure to certain thermal conditions, such as changes in the set points for the onset of sweating, core temperature set points, and salt concentration in sweat. These adaptations might be relevant for accurately predicting comfort levels over extended periods (58–60).

These considerations have led to significant research developments. In the 1970s, **Michael Humphreys and Fergus Nicol** pioneered studies correlating outdoor climate parameters, like monthly mean outdoor air temperature, with optimal indoor operative temperatures (55,61). This work was expanded by **Richard deDear and Gail Brager** through the ASHRAE RP-884 research project, which utilized an extensive global database to correlate outdoor air temperatures with the New Effective Temperature (ET*) for indoor settings (57,58). This approach helped to refine the understanding of thermal comfort in a diverse range of building types and climatic conditions across four continents. The database used in these studies contained over 22,000 data entries from more than 160 buildings, capturing a wide array of indoor climate perceptions and environmental measurements. This database has grown to include over 100,000 entries, contributing to the ASHRAE global database of thermal comfort field measurements(62), which continues to inform adaptive comfort models and national comfort standards.

9.2.1 Data Bases for the Empirical Models

When leveraging data-driven approaches in the field of thermal comfort, it is essential to possess a **deep understanding of the underlying datasets** to fully appreciate the constraints and potential biases in model applicability. This understanding is particularly important in comfort assessment, where various building characteristics or usage conditions may indirectly influence the perceived comfort levels reported by occupants. The foundational datasets that have significantly shaped contemporary comfort assessment models are as follows:

– **CLASSIC Database:** Originated by Fergus Nicol and Michael Humphreys, this dataset includes data from field studies in air-conditioned office buildings primarily in the UK. It provides valuable insights into the thermal comfort perceptions specific to controlled indoor environments (61,63,64).

- **SCATS (Steady-State and Transient Ambient Comfort Studies):** Focused on thermal comfort within transient environments, SCATS has been instrumental in developing the European adaptive comfort standards such as EN 15251 and ISSO 74 based on five different European countries. This dataset explores how rapid changes in environmental conditions affect occupant comfort (61,63,64).
- **ASHRAE RP-884 (ASHRAE Global Thermal Comfort Database I):** This extensive collection began as a compilation of data from field studies conducted in both air-conditioned and naturally ventilated buildings across various climates. Initially gathering data from studies conducted between 1982 and 1997, it covered 160 buildings worldwide, primarily commercial offices (57,58,62)
- **GTCD-II (ASHRAE Global Thermal Comfort Database II):** In 2014, the ASHRAE Global Thermal Comfort Database II (GTCD-II) project was initiated to update and expand the RP-884 database, incorporating data from the last two decades. This enhanced database, after undergoing a stringent quality assurance process, included 77,304 rows of data combining subjective comfort votes with objective measurements of thermal comfort parameters. ASHRAE Global Comfort Database II is a valuable resource for professionals and researchers interested in thermal comfort (62).

These datasets provide a robust platform for assessing thermal comfort, but initial analyses revealed that linear correlations between outdoor air temperature and operative indoor temperature in free-running buildings were influenced by several additional factors. These include building usage type, air velocity, and air humidity. Systematic deviations observed in these relationships underscore the complexity of thermal comfort and the need for models that adapt to a variety of environmental and personal factors.

9.2.2 A deeper look at the ASHRAE Global Comfort Database

The ASHRAE Global Comfort Database, which underpins some of the adaptive models explained in the next section (Section 9.2.3), is a comprehensive resource that aggregates a vast array of data from various climates, seasons, and building types, providing a solid foundation for many of these empirical models. It includes contributions from multiple continents, with a significant portion of the data derived from Europe and covers a range of building types and user environments. An overview of the composition of the current ASHRAE global comfort database is provided in (62), with *key characteristics* as follows:

- **Continents and Countries:** Approximately one-third of the datasets originated from Europe, another third from Asia, and the remaining third from various locations including South and North America, Africa, and Australia. Notably, the UK contributed most European datasets, accounting for about 25,000 out of 33,000.

Other European countries such as Slovakia, Germany, Denmark, Greece, Italy, Portugal, Sweden, Belgium, and France also made valuable contributions.

– *Climate Zones:* About one-third of the datasets pertain to Hot-Summer and Warm-Summer Mediterranean Climates. Additionally, one-twentieth of the datasets cover Central European Temperate Climate.

– *Seasons:* The database encompasses all seasons, with a specific focus on winter and summer, each accounting for one-third of the datasets. Data for spring and autumn seasons are also available, providing a comprehensive view of thermal comfort throughout the year.

– *Conditioning Types:* The database encompasses a range of conditioning types, comprising air-conditioned, mixed-mode conditioned, and naturally ventilated buildings. Among European datasets, there is an approximately equal distribution between naturally ventilated and air-conditioned buildings, each accounting for about 50 percent. A minor portion of the European data pertains to mixed mode or mechanically ventilated buildings.

– *Building Characteristics:* Detailed information about building characteristics, such as construction materials, and window types, is partially included.

– *User Types and User properties:* The database includes residential buildings, commercial spaces like offices, educational institutions such as schools and universities, healthcare facilities, and public buildings like libraries and government offices. A predominant focus lies on office spaces, especially for the European datasets as illustrated in Figure 9.1.

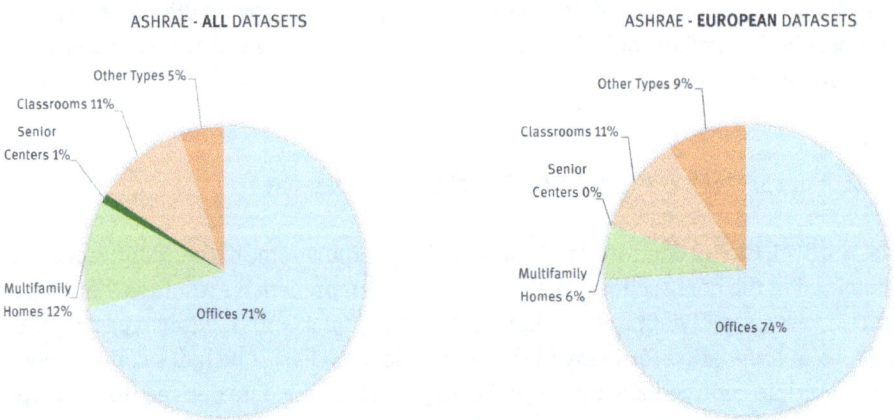

Fig. 9.1: Building Types included in the ASHRAE Global Comfort Database II, based on (65)

In conclusion, the ASHRAE Global Comfort Database is a valuable resource for professionals and researchers interested in thermal comfort. With a predominant focus on office spaces, it offers detailed insights into the impact of building characteristics, climate conditions, and user profiles on thermal comfort.

9.2.3 Empirical Equations and Their Applications

Initial assessments of these databases involved calculating linear correlations, including increments and offsets, between an outdoor air temperature characteristic and the operative room temperature for all free-running buildings. Subsequent analyses, incorporating expanded data, revealed that additional factors influence this linear relationship. Systematic deviations were observed based on building usage type, air velocity, and air humidity. This variability necessitates a multitude of models, as depicted in Table 9.1, which have been partially integrated into national and international standards, reflecting their adjustments across diverse environmental settings.

These **models differ primarily in three aspects**. First, the **slope**, where a higher factor correlates the outdoor air temperature characteristic more strongly with the optimal room temperature. Second, the **offset value**, which indicates how much the room air temperature and outdoor air temperature are shifted against each other under neural environmental conditions. Third, the **type of outdoor air temperature characteristic** used, which determines the degree of smoothing applied to the outdoor temperature in the calculation. In relation to the mentioned climatic boundary conditions (humidity, air velocity), it can be assumed that the slope value is greater the stronger the local characteristic values of these conditions negatively impact the thermal balance of the occupants.

A general model for all US-locations is provided by the equation from ASHRAE 55 (2020) suggests that the optimal temperature (θ_{opt}) can be predicted by adding 17.8 °C to 31% of the mean daily temperature ($\theta_{day,mean}$). This model applies universally across different locations and assumes buildings are free-running, meaning they are not using mechanical ventilation or cooling. The models from the European Standards EN 15251 (2007) and ISSO 74 (2014) are similar, both recommending adding 18.8 °C to 33% of the running mean air temperature outside ($\theta_{r,mean}$). EN 16798-1 (2015) proposes a slightly different formula, suggesting adding 18.0 °C to 25% of the running mean outdoor air temperature. The Chinese standard GB/T 50785 (2012) provides different formulas based on climate specifics. For hot locations, a significant offset of 9.34 °C plus 77% of the mean room temperature suggests a higher tolerance for warmth. In contrast, for mild locations, a small negative offset indicates a cooler indoor temperature preference. Additional research-driven models differentiate between dry and humid hot conditions, which is crucial for regions where humidity significantly impacts thermal comfort.

Tab. 9.1: Summary of Empirical Equations and Their Applications.

Model Restrictions	Equation	Source
All locations, free running	$\theta_{opt} = 17.8 + 0.31 \cdot \theta_{day,mean}$	ASHRAE 55 (2020)
All locations, free running	$\theta_{opt} = 18.8 + 0.33 \cdot \theta_{r,mean}$	EN 15251 (2007)
All locations, summer, [A]	$\theta_{opt} = 18.8 + 0.33 \cdot \theta_{r,mean}$	ISSO 74 (2014)
All Locations, free running	$\theta_{opt} = 18.0 + 0.25 \cdot \theta_{r,mean}$	EN 16798-1 (2015)
Hot locations, specifications [B]	$\theta_{opt} = 9.34 + 0.77 \cdot \theta_{r,mean}$	GB/T 50785 (2012)
Mild locations, specifications [B]	$\theta_{opt} = -0.31 + 0.87 \cdot \theta_{r,mean}$	GB/T 50785 (2012)
Hot dry locations, free running	$\theta_{opt} = 14.0 + 0.58 \cdot \theta_{out}$	(66)
Hot humid locations, free running	$\theta_{opt} = 12.9 + 0.57 \cdot \theta_{out}$	(66)

[A] The application requirements are clearly defined and require adaptability of the clothing and moderate activity. A correction factor is applied if this is not the case (67).
[B] The Chinese Standard includes distinction of climate zones and room climate categories, both resulting in varying slopes and offsets for the correlation function (68).

These models illustrate the complexity and variability in defining optimal indoor temperatures in adaptive model approaches, influenced by external temperatures, building characteristics, and human activity. Adaptive thermal comfort models have ushered in a nuanced approach to building design, striving to accommodate the varied thermal preferences of individuals influenced by both personal and cultural factors. However, their implementation comes with its own set of challenges. The *adaptation of these models* into mainstream standards and practices is *hindered by their simplicity*. While models like ASHRAE Standard 55 and GB/T 50785 have begun to integrate adaptive elements, there remains a gap in their universal application due to the varied individual and cultural responses to thermal conditions and their mutual influence. Adaptive models are limited in extreme climates where dependency on external factors such as the regional climate and specific building designs is high. Their effectiveness can be compromised in environments that deviate significantly from the 'norms' where the data was initially collected.

The *development and refinement* of adaptive models are *heavily reliant on comprehensive data collection* from diverse real-world environments. Initiatives like the ASHRAE Global Thermal Comfort Database II represent significant progress, yet the continuous expansion of these databases is crucial. The advent of advanced data mining and machine learning techniques presents new opportunities to enhance these models' predictive power and applicability. Ongoing research is needed to address the challenges of applying adaptive models across different building types and climatic conditions.

9.3 Remark on Combined Models

The limitations of both heat balance-based and empirical model approaches have been well-documented. Criticisms extend beyond the models themselves to include the assumptions regarding boundary conditions. For heat balance-based models, a significant issue is the disparity between the laboratory conditions under which climate chamber studies are conducted and the broad range of real-world individual and climatic conditions (69–71). These controlled laboratory environments fail to account for the variability and complexity of actual living spaces and occupant behaviors. The models' limitations are further compounded by challenges associated with the application and interpretation of determined thermal assessment scales. The PMV model overestimates heat sensation and underestimates cold, leading to inaccuracies, especially in hot climates as explained in section 9.1 In contrast, empirical models, while offering a more adaptive approach, are often hindered by their simplicity. These models typically rely on linear correlations between outdoor and indoor temperatures but do not adequately capture the dynamic interactions between occupants and their environments. Both model types also face difficulties related to the interpretation of thermal comfort scales, which depend on subjective self-reporting. This can lead to underestimation of discomfort in high temperatures.

To address the discrepancies between predicted and observed thermal comfort, several adaptive thermal balance models have been developed, incorporating adaptation mechanisms into thermal balance corrections:

– **ePMV (Extended Predicted Mean Vote):** Developed by Fanger and Toftum, the ePMV model adjusts the traditional PMV by reducing the metabolic heat input and introducing an expectancy factor. This adjustment aims to make the model more applicable to non-air-conditioned buildings in warm climates where occupants are assumed to reduce their activity with rising ambient temperatures. Data from the ASHRAE RP-884 field study, including observations from several summer-hot locations, were used. Metabolic rates were reduced by 6.7% for every thermal sensation scale unit above neutral, and an expectancy factor between 0.5 and 1.0 was applied based on the prevalence of air conditioning in the region (72).

– **aPMV (Adaptive Predicted Mean Vote):** Yao et al.'s model introduces an adaptive feedback coefficient (λ) to adjust the PMV based on the difference between predicted and observed comfort. This model relies on extensive field surveys and environmental monitoring conducted at Chongqing University, China. The adaptive coefficient λ is positive in warm environments and negative in cold ones, reflecting the degree of behavioral and psychological adaptation, leading to a reduced overestimation for positive values on the PMV scale (73).

– **aSET (Adaptive Standard Effective Temperature):** Gao et al. applied adaptive concepts to the Standard Effective Temperature (SET) index, incorporating the effects of elevated air velocity on thermal sensation. Data from a university campus in Xi'an, China, were used, where the adaptive coefficient λ varied significantly from other regions, demonstrating the influence of local climatic conditions on thermal adaptation (74).

- **ATHB (Adaptive Thermal Heat Balance):** Schweiker and Wagner's model incorporates behavioral adaptation through clothing insulation and metabolic rate adjustments to quantify both physiological and psychological adaptive processes. This model combines the adaptive comfort approach with existing PMV heat balance model by setting up equations for each of the three adaptive processes—behavioural, physiological, and psychological—to modify the mentioned input values for the PMV calculation. The coefficients for these equations are derived from empirical data and subsets of the ASHRAE RP884 database (75).

In summary, while none of the adapted heat balance-based models provide a comprehensive and compelling evidence-based explanation for adaptive comfort models, they do make incremental contributions to the domain. The reception of these adapted models in the community has been mixed, with cautious acceptance and calls for further validation. However, the increased complexity of these extended heat balance models contradicts the numerous degrees of freedom available to building occupants in naturally ventilated buildings, which influence all adaptation mechanisms (behavioral, psychological, and physiological).

The ATHB model, as introduced by Schweiker and Wagner, is considered the most advanced approach, incorporating behavioral, physiological, and psychological adaptations (64). Despite this progress, the persistent discrepancy between predictions of these adapted models and empirical observations in naturally ventilated buildings in warm climates underscores the need for a rational, evidence-based parameterization of the psychological adaptive concept of expectation.

10 Conclusion

This chapter has focused on the suitability of current insights for defining tolerance thresholds (bearability) and comfort zones for buildings during the summer. The term 'bearability' has been introduced with the hope that the gathered insights will contribute to more binding regulations for buildings in summer conditions, particularly in Europe. Additionally, by explaining the various comfort models, the chapter aims to consolidate the potentials and limitations of existing approaches.

In terms of **bearability limits**, which aim to ensure indoor climates that pose no risks for occupants, this chapter proposed a novel methodology based on climate chamber studies to establish explicit threshold values for healthy adults. Additionally, it emphasises the heightened sensitivity of specific demographic groups such as the elderly, individuals with pre-existing conditions, and young children. By deriving threshold values from excess mortality research during heatwave periods in Europe, the chapter highlights an underexplored area crucial for establishing minimum health requirements for indoor climates.

Regarding the current **state of research on comfort models** and their suitability for assessing summer comfort, the chapter reveals significant limitations of the Fanger-PMV model in determining optimal summer indoor air temperatures, whether in air-conditioned or non-air-conditioned environments. Instead, empirical approaches appear more appropriate, although their effectiveness heavily depends on the relevance and applicability of the data used. For European studies, the predominant focus on office buildings underscores a gap in the applicability of findings to other building types, such as residential homes, necessitating further investigation. Additionally, the limitations of both empirical and heat balance-based models are discussed, noting that they often fail to account for nuances associated with usage. The increased complexity of extended heat balance models contrasts with the numerous degrees of freedom available to building occupants in naturally ventilated buildings, which influence all adaptation mechanisms (behavioural, psychological, and physiological). This complexity underscores the ongoing debate between traditional heat balance approaches and newer adaptive strategies, highlighting the dynamic nature of research in this area. The potential for adaptive thermal comfort models to contribute meaningfully to building design and occupant well-being is vast, urging continuous advancements in the field.

11 References

1. **ASHRAE** (American Society of Heating, Refrigerating and Air-Conditioning Engineers). *ASHRAE 55:2023 Thermal Environmental Conditions for Human Occupancy* [Internet]. ASHRAE 55:2023 2023. Available from: https://www.ashrae.org/technical-resources/bookstore/standard-55-thermal-environmental-conditions-for-human-occupancy.
2. **ASR A3.6.** *Technische Regeln für Arbeitsstätten - Lüftung (ASR A3.6)* [Internet]. Ausschuss für Arbeitsstätten -ASTA-Geschäftsführung -BAuA; 2012. Available from: http://www.baua.de.
3. **Ekberg, L.** *Revised Swedish Guidelines for the Specification of Indoor Climate Requirements released by SWEDVAC.* 2007 [cited 2024 Apr 27];
4. The Chartered Institution of Building Services Engineers **(CIBSE)**. *TM52: The Limits of Thermal Comfort: Avoiding Overheating in European Buildings* [Internet]. TM52:2013 Oct, 2013 p. 24. Available from: https://www.cibse.org/knowledge-research/knowledge-portal/tm52-the-limits-of-thermal-comfort-avoiding-overheating-in-european-buildings.
5. The Chartered Institution of Building Services Engineers **(CIBSE)**. *TM59 Design methodology for the assessment of overheating risk in homes* [Internet]. TM59:2017 May, 2017 p. 17. Available from: https://www.cibse.org/knowledge-research/knowledge-portal/technical-memorandum-59-design-methodology-for-the-assessment-of-overheating-risk-in-homes.
6. **O'Brien W, Tahmasebi F,** editors. *Occupant-Centric Simulation-Aided Building Design: Theory, Application, and Case Studies.* New York: Routledge; 2023. Available from: https://doi.org/10.1201/9781003176985.
7. **IEA.** *The Future of Cooling - Analysis* [Internet]. 2018 [cited 2024 Jun 14]. Available from: https://www.iea.org/reports/the-future-of-cooling.
8. **Schild K.** *Mindestwärmeschutz. Wärmebrücken* [Internet]. Wiesbaden: Springer Fachmedien Wiesbaden; 2018 [cited 2024 Jun 14]. p. 19-34. Available from: http://link.springer.com/10.1007/978-3-658-20709-0_3.
9. **Russo S, Sillmann J, Fischer EM.** *Top ten European heatwaves since 1950 and their occurrence in the coming decades.* Environ Res Lett [Internet]. 2015 [cited 2023 Aug 9];10:124003. doi: 10.1088/1748-9326/10/12/124003.
10. **Oliveira A, Lopes A, Correia E.** *Annual summaries dataset of Heatwaves in Europe, as defined by the Excess Heat Factor.* Data in Brief [Internet]. 2022 [cited 2024 Apr 27];44:108511. doi: 10.1016/j.dib.2022.108511.
11. **Revich B, Shaposhnikov D.** *Excess mortality during heat waves and cold spells in Moscow, Russia.* Occupational and Environmental Medicine [Internet]. 2008 [cited 2023 Aug 15];65:691-696. doi: 10.1136/oem.2007.033944.
12. **Campbell S, Remenyi TA, White CJ, Johnston FH.** *Heatwave and health impact research: A global review.* Health & Place [Internet]. 2018 [cited 2023 Aug 15];53:210-218. doi: 10.1016/j.healthplace.2018.08.017.
13. **Arsad FS, Hod R, Ahmad N, Ismail R, Mohamed N, Baharom M, Osman Y, Radi MFM, Tangang F.** *The Impact of Heatwaves on Mortality and Morbidity and the Associated Vulnerability Factors: A Systematic Review.* Int J Environ Res Public Health [Internet]. 2022 [cited 2023 Aug 15];19:16356. doi: 10.3390/ijerph192316356.
14. **Smargiassi A, Goldberg MS, Plante C, Fournier M, Baudouin Y, Kosatsky T.** *Variation of daily warm season mortality as a function of micro-urban heat islands.* J Epidemiol Community Health [Internet]. 2009 [cited 2023 Aug 15];63:659-664. doi: 10.1136/jech.2008.078147.
15. **Díaz J, Jordán A, García R, López C, Alberdi J, Hernández E, Otero A.** *Heat waves in Madrid 1986-1997: effects on the health of the elderly.* IAOEH [Internet]. 2002 [cited 2023 Aug 16];75:163-170. doi: 10.1007/s00420-001-0290-4.

16. **Bogdanović D, Milosević Z, Lazarević K, Dolićanin Z, Randelović D, Bogdanović S.** *The Impact of the July 2007 Heat Wave on Daily Mortality in Belgrade, Serbia.* Cent Eur J Public Health. 2013;21:140-145. doi: 10.21101/cejph.a3840.
17. **Basu R, Ostro BD.** *A multicounty analysis identifying the populations vulnerable to mortality associated with high ambient temperature in California.* Am J Epidemiol. 2008;168:632-637. doi: 10.1093/aje/kwn170.
18. **Conti S, Meli P, Minelli G, Solimini R, Toccaceli V, Vichi M, Beltrano C, Perini L.** *Epidemiologic study of mortality during the Summer 2003 heat wave in Italy.* Environmental Research [Internet]. 2005 [cited 2023 Aug 16];98:390-399. doi: 10.1016/j.envres.2004.10.009.
19. **Parsons K.** *Human Thermal Environments: The Effects of Hot, Moderate, and Cold Environments on Human Health, Comfort, and Performance,* Third Edition. 3rd ed. Boca Raton: CRC Press; 2014. Available from: https://doi.org/10.1201/b16750 .
20. **Havenith G, Fiala D.** *Thermal Indices and Thermophysiological Modeling for Heat Stress.* Compr Physiol. 2015;6:255-302. doi: 10.1002/cphy.c140051.
21. **Malchaire J, Piette A, Kampmann B, Mehnert P, Gebhardt H, Havenith G, Den Hartog E, Holmer I, Parsons K, Alfano G, et al.** *Development and validation of the predicted heat strain model.* Ann Occup Hyg. 2001;45:123-135.
22. Technical Committee ISO/TC 159/SC 5. **ISO 7933:2023** [Internet]. [cited 2024 Jun 18]. Available from: https://www.iso.org/standard/78240.html.
23. **Du C, Li B, Li Y, Xu M, Yao R.** *Modification of the Predicted Heat Strain (PHS) model in predicting human thermal responses for Chinese workers in hot environments.* Building and Environment [Internet]. 2019 [cited 2024 Jun 18];165:106349. doi: 10.1016/j.buildenv.2019.106349.
24. **Wang F, Chuansi G, Kalev K, Ingvar H.** *The Predicted Heat Strain Model (ISO7933) Severely Over- or Underestimated Core and Skin Temperature in Protective and Light Summer Clothing.* Lund University [Internet]. [cited 2024 Jun 18]. Available from: https://www.lunduniversity.lu.se/lup/publication/11d98eb6-14de-49bf-862e-fdbaf45a442d.
25. **Lazaro P, Momayez M.** *Development of a modified predicted heat strain model for hot work environments.* International Journal of Mining Science and Technology [Internet]. 2020 [cited 2024 Jun 18];30:477-481. doi: 10.1016/j.ijmst.2020.05.009.
26. **Kuehn T, Ramsey J, Threlkeld J.** *Thermal Environmental Engineering.* 3rd ed. Upper Saddle River, NJ: Pearson; 1998.
27. **Wolf ST, Cottle RM, Vecellio DJ, Kenney WL.** *Critical environmental limits for young, healthy adults (PSU HEAT Project).* Journal of Applied Physiology [Internet]. 2022 [cited 2023 Aug 25];132:327-333. doi: 10.1152/japplphysiol.00737.2021.
28. **Vecellio DJ, Wolf ST, Cottle RM, Kenney WL.** *Evaluating the 35°C wet-bulb temperature adaptability threshold for young, healthy subjects (PSU HEAT Project).* Journal of Applied Physiology [Internet]. 2022 [cited 2023 Aug 17];132:340-345. doi: 10.1152/japplphysiol.00738.2021.
29. **Millyard A, Layden JD, Pyne DB, Edwards AM, Bloxham SR.** *Impairments to Thermoregulation in the Elderly During Heat Exposure Events.* Gerontology and Geriatric Medicine [Internet]. 2020 [cited 2023 Aug 17];6:2333721420932432. doi: 10.1177/2333721420932432.
30. **Kaltsatou A, Kenny GP, Flouris AD.** *The Impact of Heat Waves on Mortality among the Elderly: A Mini Systematic Review.* J Geriatr Med Gerontol. 2018;4. doi: 10.23937/2469-5858/1510053.
31. **Schifano P, Leone M, De Sario M, de'Donato F, Bargagli AM, D'Ippoliti D, Marino C, Michelozzi P.** *Changes in the effects of heat on mortality among the elderly from 1998-2010: results from a multicenter time series study in Italy.* Environ Health. 2012;11:58. doi: 10.1186/1476-069X-11-58.
32. **Meade RD, Akerman AP, Notley SR, McGinn R, Poirier P, Gosselin P, Kenny GP.** *Physiological factors characterizing heat-vulnerable older adults: A narrative review.* Environment International [Internet]. 2020 [cited 2023 Aug 15];144:105909. doi: 10.1016/j.envint.2020.105909.

33. **Huang J, Wang J, Yu W.** *The Lag Effects and Vulnerabilities of Temperature Effects on Cardiovascular Disease Mortality in a Subtropical Climate Zone in China.* International Journal of Environmental Research and Public Health [Internet]. 2014 [cited 2023 Aug 15];11:3982-3994. doi: 10.3390/ijerph110403982.

34. **Zhang S, Rai M, Matthies-Wiesler F, Breitner-Busch S, Stafoggia M, Donato Fde, Agewall S, Atar D, Moman A M, Peters A,** et al. *Climate change and cardiovascular disease - the impact of heat and heat-health action plans.* E-Journal of Cardiology Practice. 2022;Volume 22.

35. **Hajat S, Kovats RS, Lachowycz K.** *Heat-related and cold-related deaths in England and Wales: who is at risk?* Occup Environ Med. 2007;64:93-100. doi: 10.1136/oem.2006.029017.

36. **Basu R, Ostro BD.** *A multicounty analysis identifying the populations vulnerable to mortality associated with high ambient temperature in California.* Am J Epidemiol. 2008;168:632-637. doi: 10.1093/aje/kwn170.

37. **Helldén D, Andersson C, Nilsson M, Ebi KL, Friberg P, Alfvén T.** *Climate change and child health: a scoping review and an expanded conceptual framework.* The Lancet Planetary Health [Internet]. 2021 [cited 2023 Aug 15];5:e164-e175. doi: 10.1016/S2542-5196(20)30274-6.

38. **Strand LB, Barnett AG, Tong S.** *Maternal exposure to ambient temperature and the risks of preterm birth and stillbirth in Brisbane, Australia.* Am J Epidemiol. 2012;175:99-107. doi: 10.1093/aje/kwr404. Cited: in: : PMID: 22167749.

39. **Basu R, Malig B, Ostro B.** *High ambient temperature and the risk of preterm delivery.* Am J Epidemiol. 2010;172:1108-1117. doi: 10.1093/aje/kwq170. Cited: in: : PMID: 20889619.

40. **Auger N, Naimi AI, Smargiassi A, Lo E, Kosatsky T.** *Extreme heat and risk of early delivery among preterm and term pregnancies.* Epidemiology. 2014;25:344-350. doi: 10.1097/EDE.0000000000000074.

41. **Wang Y-Y, Li Q, Guo Y, Zhou H, Wang Q-M, Shen H-P, Zhang Y-P, Yan D-H, Li S, Chen G,** et al. *Ambient temperature and the risk of preterm birth: A national birth cohort study in the mainland China.* Environment International [Internet]. 2020 [cited 2023 Aug 17];142:105851. doi: 10.1016/j.envint.2020.105851.

42. **DIN EN 16798-1:2022-03,** *Energetische Bewertung von Gebäuden- Lüftung von Gebäuden-Teil 1: Eingangsparameter für das Innenraumklima zur Auslegung und Bewertung der Energieeffizienz von Gebäuden bezüglich Raumluftqualität, Temperatur, Licht und Akustik - Modul M1-6; Deutsche Fassung EN16798-1:2019* [Internet]. Beuth Verlag GmbH; [cited 2024 Jun 18]. Available from: https://www.beuth.de/de/-/-/349622591.

43. **Humboldt, von A.** *Fragmente einer Geologie und Klimatologie Asiens* [Internet]. Berlin: J. A. List; 1832 [cited 2023 Aug 31]. Available from: https://www.digitale-sammlungen.de/de/view/bsb10433527?

44. **Gagge AP.** *Rational temperature indices of thermal comfort.* Bioengineering, Thermal Physiology and Comfort. Amsterdam: Elsevier; 1981. p. 79-98.

45. **Belding HS, Hatch T.** *Index for evaluating Heat Stress in Terms of resulting Physiological Strains.* Heating, piping, and air conditioning. 1955;

46. **Beshir M, Ramsey JD.** *Heat stress indices: A review paper.* International Journal of Industrial Ergonomics [Internet]. 1988 [cited 2023 Aug 31];3:89-102. doi: 10.1016/0169-8141(88)90012-1.

47. **Dukes-Dobos FN, Francis N, Henschel J.** *Occupational exposure to hot environments; criteria for a recommended standard.* National Institute for Occupational Safety and Health, Division of Standards Development and Technology Transfer; 1986.

48. **Fanger PO.** *Thermal comfort. Analysis and applications in environmental engineering.* Copenhagen: Danish Technical Press; 1970.

49. **Fanger PO.** *Assessment of man's thermal comfort in practice.* British Journal of Industrial Medicine. 1973;30:313. doi: 10.1136/oem.30.4.313.

50. **ISO 7730:2023** [Internet]. Geneve: International Organization for Standardization, 2023; 2023. Available from: https://www.iso.org/standard/14566.html

51. **ISO 7730:1984** [Internet]. Geneve: International Organization for Standardization, 1984; 1984 [cited 2024 Jun 18]. Available from: https://www.iso.org/standard/14566.html.

52. **Avelino A, Silva L.** *PMV as a thermal evaluation method for air-conditioned spaces in hot climates: a systematic review.* Ciência e Natura. 2020;42:e29. doi: 10.5902/2179460X41375.

53. **Cheung T, Schiavon S, Parkinson T, Li P, Brager G.** *Analysis of the accuracy on PMV - PPD model using the ASHRAE Global Thermal Comfort Database II.* Building and Environment [Internet]. 2019 [cited 2024 Jun 17];153:205-217. doi: 10.1016/j.buildenv.2019.01.055.

54. **Schweiker M, André M, Al-Atrash F, Al-Khatri H, Alprianti RR, Alsaad H, Amin R, Ampatzi E, Arsano AY, Azar E, et al.** *Evaluating assumptions of scales for subjective assessment of thermal environments - Do laypersons perceive them the way, we researchers believe?* Energy and Buildings [Internet]. 2020 [cited 2024 Jun 16];211:109761. doi: 10.1016/j.enbuild.2020.109761.

55. **Nicol JF, Humphreys MA.** *Thermal comfort as part of a self-regulating system.* Building Research and Practice [Internet]. 1973 [cited 2023 Sep 26];1:174-179. doi: 10.1080/09613217308550237.

56. **Humphreys M.** *Outdoor temperatures and comfort indoors.* Batiment International, Building Research and Practice [Internet]. 1978 [cited 2023 Sep 26];6:92-92. doi: 10.1080/09613217808550656.

57. **de Dear R, Brager GS.** *Developing an adaptive model of thermal comfort and preference.* 1998 [cited 2024 Jun 17]; Available from: https://escholarship.org/uc/item/4qq2p9c6.

58. **de Dear R, Schiller Brager G.** *The adaptive model of thermal comfort and energy conservation in the built environment.* Int J Biometeorol [Internet]. 2001 [cited 2023 Sep 14];45:100-108. doi: 10.1007/s004840100093.

59. **Zhang F, de Dear R.** *Impacts of demographic, contextual and interaction effects on thermal sensation—Evidence from a global database.* Building and Environment [Internet]. 2019 [cited 2023 Oct 20];162:106286. doi: 10.1016/j.buildenv.2019.106286.

60. **Parsons KC.** *The effects of gender, acclimation state, the opportunity to adjust clothing and physical disability on requirements for thermal comfort.* Energy and Buildings. 2002;34:593-599. doi: 10.1016/S0378-7788(02)00009-9.

61. **Humphreys MA, Nicol JF, Raja IA.** *Field Studies of Indoor Thermal Comfort and the Progress of the Adaptive Approach.* Advances in Building Energy Research [Internet]. 2007 [cited 2024 Jun 17]; Available from: https://doi.org/10.1080/17512549.2007.9687269

62. **Földváry V, Cheung CT, Zhang H, de Dear R, Parkinson T, Arens E, Chun C, Luo M, Brager G, Li P, et al.** *Development of the ASHRAE Global Thermal Comfort Database II.* Building and Environment. 2018;142. doi: 10.1016/j.buildenv.2018.06.022.

63. **Carlucci S, Bai L, de Dear R, Yang L.** *Review of adaptive thermal comfort models in built environmental regulatory documents.* Building and Environment [Internet]. 2018 [cited 2023 Sep 14];137:73-89. doi: 10.1016/j.buildenv.2018.03.053.

64. **de Dear R, Xiong J, Kim J, Cao B.** *A review of adaptive thermal comfort research since 1998.* Energy and Buildings [Internet]. 2020 [cited 2024 Jun 17];214:109893. doi: 10.1016/j.enbuild.2020.109893.

65. **Parkinson T, Tartarini F, Földváry Ličina V, Cheung T, Zhang H, De Dear R, Li P, Arens E, Chun C, Schiavon S, et al.** *ASHRAE global database of thermal comfort field measurements* [Internet]. Dryad; 2018 [cited 2023 Sep 18]. p. 4167577 bytes. Available from: https://datadryad.org/stash/dataset/doi:10.6078/D1F671.

66. **Toe D, Toe C, Kubota T.** *Reanalysing the ASHRAE RP-884 Database to Determine Thermal Comfort Criteria for Naturally Ventilated Buildings in Hot-Humid Climate.* 2012. Conference:

PLEA2012 - 28th Conference, Opportunities, Limits & Needs Towards an environmentally responsible architecture Lima, Perú. Available from: Research Gate

67. **Boerstra AC, Van Hoof J, Van Weele AM.** *A new hybrid thermal comfort guideline for the Netherlands: Background and development.* Architectural Science Review. 2015;58:24-34. doi: 10.1080/00038628.2014.971702.

68. **GB/T 50785** *Evaluation standard for indoor thermal environment in civil buildings.* 2012. Available from: https://www.chinesestandard.net/PDF/English.aspx/GBT50785-2012

69. **Nicol F. Thermal Comfort**: *A Handbook for Field Studies Towards an Adaptive Model.* University of East London; 1993.

70. **de Dear RJ, Brager GS.** *Thermal comfort in naturally ventilated buildings: revisions to ASHRAE Standard 55.* Energy and Buildings [Internet]. 2002 [cited 2024 Jun 18];34:549-561. doi: 10.1016/S0378-7788(02)00005-1.

71. **Humphreys MA, Fergus Nicol J.** *The validity of ISO-PMV for predicting comfort votes in everyday thermal environments.* Energy and Buildings [Internet]. 2002 [cited 2024 Jun 18];34:667-684. doi: 10.1016/S0378-7788(02)00018-X.

72. **Ole Fanger P, Toftum J.** *Extension of the PMV model to non-air-conditioned buildings in warm climates.* Energy and Buildings [Internet]. 2002 [cited 2024 Jun 18];34:533-536. doi: 10.1016/S0378-7788(02)00003-8.

73. **Yao R, Li B, Liu J.** *A theoretical adaptive model of thermal comfort - Adaptive Predicted Mean Vote (aPMV).* Building and Environment [Internet]. 2009 [cited 2024 Jun 18];44:2089-2096. doi: 10.1016/j.buildenv.2009.02.014.

74. **Gao J, Wang Y, Wargocki P.** *Comparative analysis of modified PMV models and SET models to predict human thermal sensation in naturally ventilated buildings.* Building and Environment [Internet]. 2015 [cited 2024 Jun 18];92:200-208. doi: https://doi.org/10.1016/j.buildenv.2015.04.030.

75. **Schweiker M, Wagner A.** *A framework for an adaptive thermal heat balance model* (ATHB). Building and Environment [Internet]. 2015 [cited 2024 Jun 18];94:252-262. https://doi.org/10.1016/j.buildenv.2015.08.018

Sabine Hoffmann, Abolfazl Ganji, Peggy Freudenberg, Christoph Schünemann

Building Simulation for Overheating Risk Evaluation and Optimization

Overview of the Basic Concepts and Application Aspects of Simulation Models.

Abstract: The increasing risk of overheating in buildings, reinforced by global warming, necessitates advanced modeling techniques to realistically assess and mitigate this hazard. This chapter covers critical modeling aspects for evaluating the overheating risk in buildings, underscoring that simulation is indispensable for an accurate evaluation. Initial discussions focus on the sources of error in thermal modeling and the vital steps for plausibility checks and validation of simulation outcomes. Quality criteria for selecting appropriate software tools are then outlined, ensuring the reliability and accuracy of simulations. Subsequently, the chapter explores simulation aspects of thermal protection measures during summer. These include modeling solar heat gain, thermal inertia, solving the 1-D heat transfer equation, and incorporating latent heat storage with phase change materials. Strategies such as enhanced thermal mass and various ventilation schemes, including increased day and night, are analyzed for their effectiveness in reducing overheating. Additional considerations involve the simulation of glazing systems, including solar transmittance, secondary heat flux, and the performance of shading systems.

Keywords: Thermal Modeling, Simulation Accuracy, Heat Mitigation Strategies, Ventilation Schemes, Solar Transmittance

Sabine Hoffmann, Rheinland-Pfälzische Technische Universität Kaiserslautern-Landau, Paul-Ehrlich-Str. 14, 67663 Kaiserslautern, +49 631 205-2909, sabine.hoffmann@rptu.de

Abolfazl Ganji, Transsolar Energietechnik GmbH, Curiestraße 2, 70563 Stuttgart, +49(0)711 67976-0, ganji@transsolar.com

Peggy Freudenberg, Dresden University of Technology, Institute of Building Climatology, Zellescher Weg 17, 01062 Dresden, +49(0)351 463-35259, peggy.freudenberg@tu-dresden.de

Christoph Schünemann, Leibnitz Institute of Ecological Urban and Regional Development (IOER), Weberplatz 1, 01217 Dresden, +49(0)351 4679-194, c.schuenemann@ioer.de

12 Quality Assurance and Error Management

Buildings are complex entities where dynamic interactions and operation processes critically influence the performance outcome. The vast parameter space and the multitude of potential measures in construction projects render analytical equations insufficient for fully capturing this complexity. Hence, Building Energy Simulation (BES) becomes indispensable, particularly for complex projects. BES leverages numerical methods to model the heat and mass transfers which occurs within and bet ween the building envelope and its HVAC systems under realistic, transient conditions. This simulation approach utilizes discretized partial differential equations across time and space, typically solved via implicit finite difference methods or their derivatives. This allows BES to accommodate everything from detailed mass flow simulations to more general cooling and heating demands. The time steps applied to different simulation models can vary from one minute for highly transient processes to one hour for cases with less dynamics. BES simulations are typically run over an entire year, therefor referred to as annual simulations, or at least an extended period of several days, weeks or months.

By advancing beyond standard metric-based approaches as usually prescribed in codes and standards, and embracing detailed, validated simulation tools, engineers and designers can enhance the accuracy and reliability of their predictions, hence ensuring that buildings meet energy efficiency standards, sustainability goals, and thermal comfort requirements. This vast potential and complexity of building simulation underscore the need for rigorous quality assurance protocols and error management strategies, especially when bridging the discrepancy between predicted and actual building performance.

12.1 Sources of Error

While physics of heat transfer are well established and the applied numerical methods in general sound, there are some disturbances in the modelling of buildings which are detrimental to the accurate simulation-based prediction of energy consumption and comfort conditions:

1) *Occupant behaviour and manual operation of systems* greatly influence building performance but are inherently difficult to predict due to their stochastic nature. These behaviours are influenced by a multitude of factors, including individual preferences, specific environmental conditions and characteristics that are not represented in the model or weather file. Academically, there is a growing effort to model these stochastic interactions within the built environment more accurately. However, for practical purposes in the design phase, simulations usually rely on predetermined schedules to account for internal loads and system

operations because more detailed operational data is lacking. These schedules can be adjusted to different day types—weekdays, Saturdays, and Sundays—to reflect typical usage patterns, but at the same time schedules oversimplify the true complexity introduced by variable human behaviour.

2) ***Weather files*** used for BES contain meteorological variables and aim to represent typical climatic conditions. They often fail to match the specific ambient conditions at a particular location, especially in urban areas where local climate can significantly diverge from regional averages. Unfortunately, spatially resolved datasets from national weather agencies are not universally available, leaving gaps in data. In such instances, the simulation of local microclimates with specialized tools (e.g., ENVI-met (1)) can provide valuable insights. This method, along with the application of specific weather files for extreme weather periods and future climate scenarios, is discussed through case studies in Chapter One.

3) ***Input data***, such as material properties, construction layers, and HVAC system controls are frequently unknown, especially in existing buildings, or subject to change during the constructions and/or operation phase. The accuracy of simulation results relies heavily on the quality of these inputs. In the context of summer conditions, a particular emphasis should be placed on the properties of window systems including solar protection, as these significantly affect building performance. Comprehensive databases for glazing and shading systems are available in simulation tools, but their effective use requires selecting the appropriate software ensuring it has sufficient accuracy to model individual systems accurately. Similarly, assumptions for ventilation rates need to be realistic and ideally air flow networks which are validated through measurement data should be employed.

These three major sources of error are inevitably inherent to some extent in each simulation model. A potential mismatch of input data and the almost erratic behaviour of human beings and weather can lead to large discrepancies between simulated energy consumption and utility bills of real buildings. This discrepancy is often called 'performance gap' and has been discussed widely in the building science community. The difference between modelled and actual behaviour of buildings led to a certain generalized scepticism amongst lay people that are not familiar with the basic principles of building simulation. Here it is necessary to clarify that building simulation is first and foremost a tool to assess relative changes related to the variation of certain parameters and schemes, while others are kept constant, e.g., the reduced or increased energy consumption related with different variants of building envelope. It will rarely predict absolute values precisely unless a detailed monitoring of the above-mentioned variables is carried out.

12.2 Plausibility Checks and Validation

In addition to the inherent potential sources of error and possible bugs in the software code, Building Energy Simulation often allows for faulty user input due to negligence or a lack of understanding of building physics. Therefore, it is crucial to conduct continuous and thorough plausibility checks with every simulation. The sheer number of inputs and the selection of calculation models in comprehensive BES tools are extensive and sometimes even overwhelming. Plausibility checks, such as setting boundary conditions to constant values and performing analytical energy balance checks before allowing dynamic conditions, are recommended. While sanity checks primarily address potential user errors, it is also necessary to validate any new models that developers wish to implement in BES. Generally, validating a numerical model within BES involves comparing the calculated outcomes with results from other approaches:

–   *Analytical tests* which provide closed form solutions for certain building dynamics. The comparison with analytical models can be used for simple cases, e.g., the temperature distribution within layers and/or components with Dirichlet or Neumann boundary conditions.

–   *Empirical tests* that are based on measured data, providing real-world scenarios that can be particularly useful for testing and validating the simulation of summer conditions.

–   *Comparative tests* which do not rely on empirical data but compare outcomes across different simulation tools to ensure consistency and accuracy. Validation of new models or calculation approaches that are implemented in more than one of the major BES tools can be carried out through cross comparison. The simulation set-ups that are used for such validation should be kept as simple as possible to avoid discrepancies due to differences in underlying models other than the one investigated.

To ensure the precision and reliability of BES models, various standards and guidelines are currently employed. These include EN ISO 13791 (2) and VDI 6007 (3), which provide fundamental test cases for the thermal performance of buildings, specifically addressing the variable behaviours of buildings across different seasons. The international guideline, ASHRAE 140 (4), also known as BESTest (Building Energy Simulation Test), offers a robust framework for the validation of energy simulation programs. The ASHRAE Standard 140 outlines procedures for testing BES tools, incorporating tests like the thermal fabric test cases which are essential for assessing the accuracy of building thermal simulations during summer conditions when the impact of fenestration properties, including glazing and shading systems, is most pronounced. VDI 6007 and VDI 6020 (5) also stress the requirements for thermal-energy-related calculation methods in building and system simulations, underscoring the critical role of data quality and model fidelity in the simulation process.

Validating simulation tools with measurement data presents several challenges that can affect the accuracy and reliability of the results. First, measurement errors from the instrumentation and procedural mistakes during data collection can introduce discrepancies. Second, deviations can occur during data evaluation, including data averaging, cleaning, and calibration processes such as climate data analysis. Furthermore, differences in model approaches and their corresponding parameter calibration methods can lead to variations in outcomes. Additionally, divergences in implementation techniques and potential programming errors in the computation core or in data input/output can complicate the validation process. Lastly, errors in result data analysis may arise from ambiguous definitions or conversion mistakes among units. Each of these factors must be carefully considered and addressed to ensure that the validation of simulation tools against empirical data is both accurate and useful.

The *SimQuality* project represented a significant initiative in the field of building simulation, focusing on enhancing the quality and reliability of simulation tools used for assessing building energy performance and indoor environmental quality (6, 7). This project addressed the critical need for *standardizing the validation processes* of simulation tools, especially those utilized for evaluating overheating risks in buildings. The project consortium, consisting of researchers and BES program developers, systematically tackled the challenges inherent in the validation of building simulation tools by establishing a comprehensive framework that integrated rigorous *test cases, validation methodologies, and clear standards* for comparing simulation outputs against empirical data. A central feature of SimQuality was the development of a unified platform (*interactive test suite*) that allowed for the interactive comparison of simulation tools across a variety of scenarios (8).

Typically, comparisons between simulation outputs and real-world data are conducted visually using graphical representations such as diagrams. This method allows for a preliminary qualitative assessment but lacks the precision of numerical validation. *Numerical comparison and numerical acceptability assessment* are less commonly implemented yet are *crucial for detailed validation*. These involve statistical metrics that quantify differences between simulated and measured data, assessing the accuracy of the simulation models on a more granular level. While considerable research has been conducted on comparing energy-related values—such as energy consumption—using these statistical indices, there is a substantial gap when it comes to other outputs, particularly temperature profiles (9). These outputs require more nuanced analysis due to their continuous nature and the complex dynamics involved, such as time delays and amplitude attenuation (10).

12.3 Application Criteria for BES-Tools

Selecting the appropriate BES tools is a critical decision influenced by various factors that reflect broader trends in the programming industry. The range of BES tools spans from freely available open-source software to sophisticated commercial offerings that require licensing fees. Critical considerations in choosing a BES tool include ease of use, performance capabilities, and the balance between these factors, which are essential for effective application in practical engineering contexts. The variety of tools available includes *EnergyPlus (11), TRNSYS (12), ESP-r (13), IDA ICE (14)*, and others, each providing extensive capabilities for detailed simulation tasks. When selecting a BES tool, several criteria should be considered to ensure it meets the specific needs of the project:

–   *Open-Source Code:* Availability of options to implement new models or modify existing modules. Open-source tools are often free, which enhances accessibility and adaptability. Additionally, the open-source nature allows for verification and validation of the models by the user community, ensuring transparency and reliability.

–   *Licensing Costs and Pricing Policy*: The costs for commercial software licenses can vary widely. Some tools offer basic versions for free, while advanced features are priced.

–   *Documentation Quality and Support:* Access to comprehensive guides and robust user support. It is crucial that the documentation not only guides the user through the features and usage of the software but also provides detailed descriptions of the underlying models, including literature references.

–   *Educational Resources:* Availability of tutorials and learning materials is essential to facilitate user understanding and effective application of the software, ensuring users can fully leverage the capabilities of the tool.

–   *User Interface, Compatibility, and Third-party Tool Integration:* A user-friendly graphical interface, support for various operating systems (Windows, Mac, Linux/Unix), and the ability to integrate with third-party tools like Radiance for lighting simulation and CFD software enhance versatility and user experience.

13 Modeling Aspects

When using Building Energy Simulation to assess the quality of thermal protection against overheating it is essential to capture the main influences on thermal indoor conditions. This section briefly reviews critical modelling aspects that affect building performance during summertime. While comprehensive resources such as the works by Joe Clarke (15) and Jan Hensen (16) provide invaluable insights for both beginners and seasoned BES practitioners, the focus here is to illuminate specific elements without extensively exploring underlying equations or serving as a detailed tutorial. The key modelling aspects for summer thermal protection covered in the next sections are:

– *Solar Radiation Gains on Facades and Roofs*: Simulation models must accurately account for the intensity and angle of solar radiation impacting building surfaces. Correctly modelling how solar radiation interacts with these surfaces is crucial for determining the resultant heat gain and, consequently, the risk of overheating.

– *Heat Transfer through Glazing and Complex Fenestration Systems*: Advanced fenestration systems, such as switchable glazing and external shading devices, play a vital role in managing the amount of solar energy that enters a building. Models need to capture the dynamic and angular-dependent behaviour of these systems effectively to evaluate their impact on reducing unwanted heat gain while allowing for natural light.

– *Heat Transfer through Opaque Walls and Thermal Inertia Modelling*: The inclusion of thermal mass significantly influences indoor temperatures by absorbing and releasing heat. Different models exist for simulating thermal mass, each with its own set of application limits. Special attention is paid to modelling enhanced thermal storage capacities through the use of phase change materials (PCMs), which are adept at regulating temperature fluctuations due to their energy storage and release capabilities.

– *Natural Ventilation Modelling*: Accurate modelling of natural ventilation requires simulating air flow networks and control strategies that optimize indoor air quality and temperature during summer. Models should include the effects of wind and buoyancy-driven flows to enhance the effectiveness of natural ventilation strategies.

13.1   Solar Radiation Gains on Facades and Roofs

Solar radiation modeling is an integral aspect of evaluating and mitigating the overheating risk in buildings. Therefore, both diffuse and direct components of solar radiation need to be derived from climate data sets. These sets originate from empirical measurements or are generated algorithmically, or from a combination of both approaches. Weather files include solar radiation data usually as hourly values of Global Horizontal Irradiation (GHI), Direct Normal Irradiation (DNI) and Diffuse Horizontal Irradiation (DHI).

The solar altitude angle represents the height of the sun above the horizon, while the azimuth describes the directional component along the horizontal plane, defined typically from north in a clockwise direction (alternatively with 0° for south, -90° for east, 90° for west and 180° for north). The maximum solar altitude angle at the summer solstice in mid-June can be directly deduced from the latitude of the location. The accurate sun position for any location specified by latitude and longitude can be determined mathematically and the Equation of Time can be used to calculate the difference between solar time and local time. It is noteworthy that BES tools usually ignore the shift to daylight saving time and this correction has to be applied onto the output data. As climate data sets typically provide hourly values and BES tools may use smaller time steps, one of the challenges in the calculation of solar radiation consists in interpolating GHI and DNI values from the weather files. This can lead to significant discrepancies between different software tools, in particular for the hours of sun set and sun rise.

The angle of incidence of the direct sun beam changes continuously over the day and the daily pattern changes over the year. Only the detailed consideration of sun position and sky conditions in building simulation allows therefor for a precise calculation of solar radiation incident on the façade. The frequency distribution of the incidence angle[1] of solar radiation for three façade orientations, simulated with Therakles, is shown in Figure 13.1. Although almost all directions of incident radiation are covered for west-, south-west, and south-facing façades over the course of the year, a clear shift towards higher angles of incidence is found for south-west and south orientation in summer. Solar incident angles of less than 50° are therefore the exception for these two orientations. The west-facing façade in contrary is dominated by low incidence angles and therefor high radiation intensities.

---

1 Incidence angle is defined as the angle between the facade normal and the sun beam.

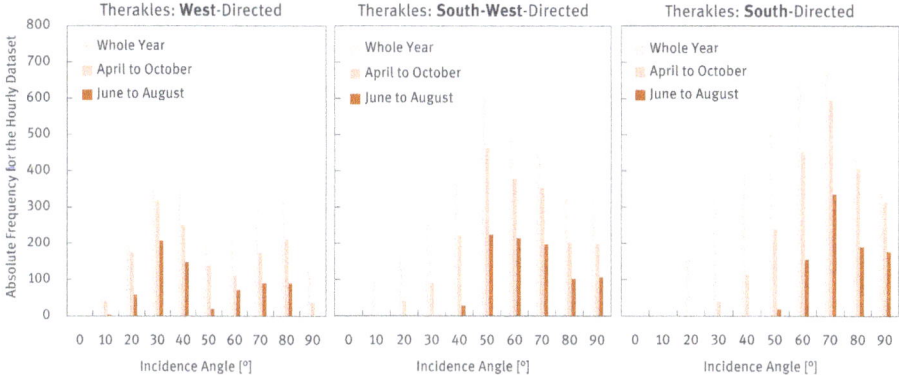

Fig. 13.1: Comparison between the frequencies (y-axis) of the incidence angle of direct solar radiation (10° ranges on x-axis) for three different façade orientations and the TRY dataset for Potsdam 2011 (latitude = 52.38 °, longitude = 13.07 °) (simulated with Therakles)

13.2 Heat Transfer through Glazing and Complex Fenestration Systems

Fenestration systems describe the transparent parts of the façade which usually consist of multiple plane insulating glazing units (IGU), either as window system or as non-operable glazed fenestration system. The term Complex Fenestration Systems (CFS) usually refers to IGUs in combination with dynamic glazing or scattering shading systems.

13.2.1 Angular Dependency of Solar Heat Gain

A key metric to evaluate the performance of a fenestration with respect to thermal protection in summer is the **Solar Heat Gain Coefficient (SHGC)** or **g-value**, which measures the fraction of incident solar radiation admitted through a window including the directly transmitted portion and the portion of the absorbed radiation which is released inward as so-called **Secondary Heat Flux**. SHGC values are typically calculated according to the ISO 15099 standard, while g-values are calculated according to EN 410. The resulting values are similar, but not identical due to slightly different boundary conditions applied in the two standards.

To accurately determine SHGC or g-values for non-measured fenestration systems, software tools such as LBNL Window are used extensively. An important feature is the integrated **International Glazing Database (IGDB)**. Managed by the Lawrence Berkeley National Laboratory, the IGDB contains detailed spectral data and angle-dependent properties for over 6,000 types of glass produced by manifold international manufacturers.

Transmittance and consequently solar heat gain coefficients are highly angular-dependent as can be seen in Figure 13.2 where different glazing systems are compared. There is a significant percentage reduction of SHGC values as the angle of incidence raises, in particular for glazing systems with reflective coatings in the near-, mid-, and/or far infrared range (solar protective coatings, low-emissivity coatings).

The strong angular dependency of solar heat gain in combination with the varying frequency of incident radiation over time and based on façade orientation (refer to Figure 13.1) underlines the necessity to run detailed transient simulations to correctly assess the actual solar heat gain into the building and the resulting thermal conditions.

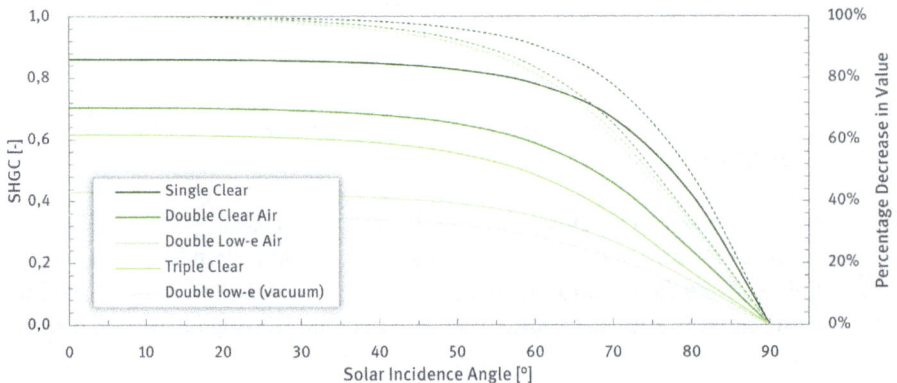

Fig. 13.2: Selection of glazing systems showing the angular SHGC on the primary Y-Axis (solid lines) and their percentage value decrease on the secondary Y-Axis (dashed lines) (modelled with LBNL Window)

13.2.2 Spectrally Selective, Thermochromic, and Electrochromic Glazing Systems

As solar heat gain is one of the determining factors for indoor conditions during summertime, many efforts have been made by glazing manufacturers to control the energy transmitted into the building. In order to represent selective and/or dynamic glazing systems correctly in BES, particular attention has to be paid to simulation input and models. Figure 13.3 shows the transmittance value of three different types of glazing as examples:

a) *Spectrally selective glazing layer with low-e coating*: While in the visible spectrum of 380 to 780 nm the glazing still transmits a significant portion of the incident radiation (here: $T_{vis} = 0.45$), it is almost opaque for NIR and MIR radiation in the range of 780 to 2500 nm. The overall solar transmittance of the glazing layer is calculated for the spectrum of 300 to 2500 nm (here: $T_{sol} = 0.65$). BES usually requires the input of T_{sol} for energy calculation, T_{vis} is a photometric metric used for lighting calculation.

b) ***Electrochromic glazing that changes tint according to active control:*** Electrochromic glazing (EC) changes its tint according to the voltage that is applied to the EC-layer. The BES model needs to reflect the control strategy, e.g., incident radiation or illuminance as control variables to choose which transmittance and consequently SHGC values are used. In the example shown in Figure 13.3, T_{vis} ranges from 0.01 to 0.68, T_{sol} ranges from 0.001 to 0.42. The use of active controls in BES makes the modelling process more complex, but choosing the right control strategy is crucial to exploit the full potential of switchable glazing. (17,18)

c) ***Thermochromic glazing that changes tint according to temperature:*** that changes tint according to temperature: Thermochromic glazing changes the transmittance according to the temperature of the glazing. While in the example in Figure 13.3 there is still a significant transmittance in the visible range at 24 °C, the glazing is almost opaque for high temperatures (here 76 °C). Some BES tools include models for thermochromic glazing where users need to input transmittance values according to temperature ranges.

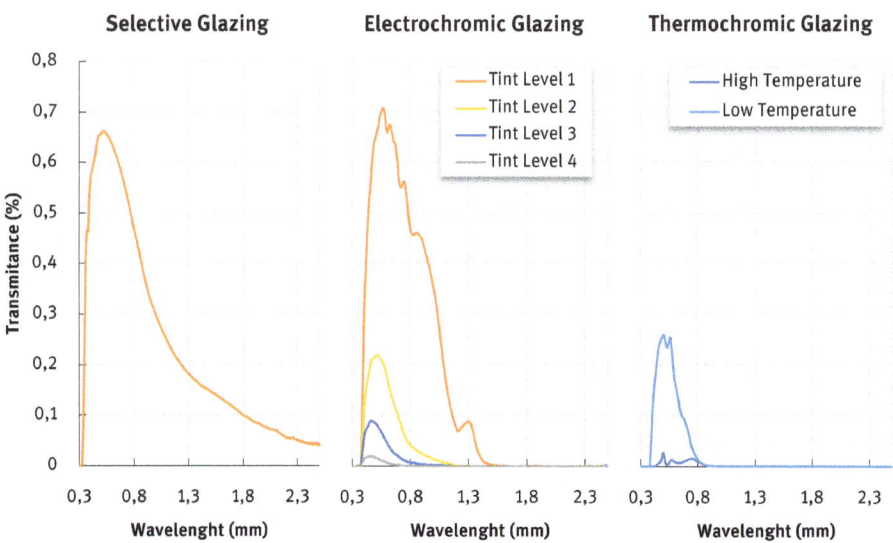

Fig. 13.3: Transmittance values of a) a spectrally selective glazing, b) an electrochromic glazing with four different levels of transmittance, c) a thermochromic glazing (Sources from IGDB, spectral measurements at Lawrence Berkeley National Laboratory, see (19))

The example shown in Figure 13.4 provides simulation results from EnergyPlus with a (fictitious) thermochromic glazing (20). It shows the frequency of how often the glazing would be partially tinted and how often it would be fully tinted dependent on incident radiation and outdoor dry-bulb temperature. The results were generated for the climate of Chicago and are of course specific to the climate data.

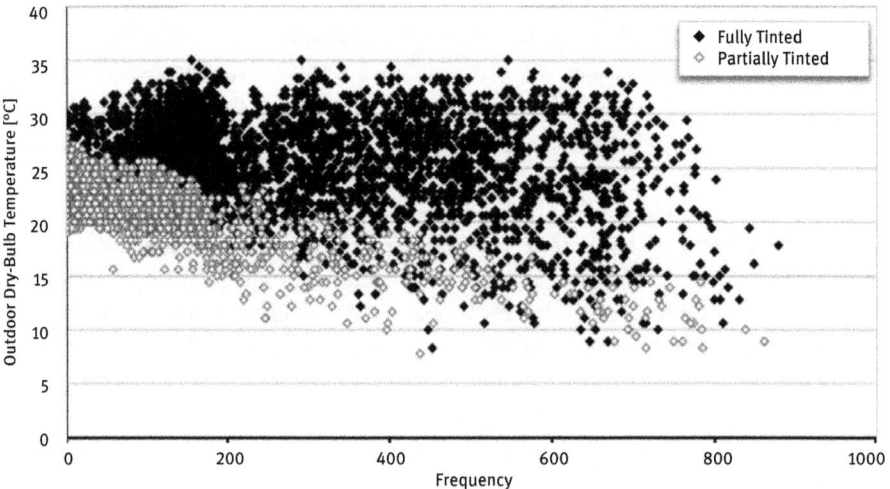

Fig. 13.4: Frequency of partial and full tint for a theoretical thermochromic glazing (20) (modelled with EnergyPlus)

13.2.3 Solar Gains through Shading Systems

In buildings with large window-to-wall ratios, i.e., with a significant portion of glazed area, external solar shading systems are well suited for preventing overheating. Most BES tools model these shading systems using a reduction factor that either decreases the solar radiation transmitted through the glazing or alters the total energy transmittance, depending on the software's capabilities.

For materials like uniformly diffuse transmitting fabrics, the use of a general reduction factor is appropriate. However, geometric shading devices such as venetian blinds, which are made from opaque materials, demonstrate a strong dependence on the angle of incidence of solar radiation. Over the past two decades, specialized software tools like LBNL Window have been developed to compute the angle-dependent properties of solar shading systems in conjunction with the glazing for various types of geometric shading systems. The incorporation of angle-dependent data into BES tools, in particular into EnergyPlus and TRNSYS 18, has been facilitated by the adoption of **bidirectional scattering distribution functions (BSDF)**.

The **BSDF methodology** involves placing two imaginary hemispheres on the façade, both internally and externally, and discretizing them. Incident direct and diffuse solar radiation is mapped to the discrete solid angles on the exterior hemi-sphere, while the radiation transmitted into the building is correspondingly mapped to the solid angles of the interior hemisphere. The solar transmittance through the façade is described by the transmission matrix ('BSDF matrix'). In this matrix, the columns correspond to the solid angles of incoming radiation from the external hem-isphere, and the rows correspond to the solid angles of radiation transmitted into the interior space. The matrix values indicate the proportion of solar radiation, from var-ious directions, that is transmitted into each of the 145 solid angles of the interior hemisphere. BSDF matrices can be tailored for different spectral ranges, including visible light or the broader solar energy spectrum, enabling a nuanced understanding of how different types of radiation are managed by shading systems.

To demonstrate the ***impact of BSDF-based shading modeling*** compared to the ***reduction factor-based approach***, comparative simulations for a venetian blind sys-tem are presented in Figure 13.5 (21). The tilt angle of the blinds was set to 45° in the studied case. In the graphs, only diffuse radiation values are shown in blue, while global radiation values (diffuse and direct radiation) are depicted in orange. While the transmitted diffuse radiation is nearly identical in both modeling variants, there are significant differences regarding the transmitted direct radiation. The BSDF method results in lower transmitted radiation values with the deviation depending on the angle of incidence of solar radiation. For incidence angles above the cut-off-angle, the blinds block almost entirely the incident direct radiation, while a value of $F_c = 0.25$ still allows 25 % of the solar heat gain.

Similarly as for glazing systems, the total solar heat gain through a complex fen-estration system (glazing plus shading) contains not only the transmitted solar radi-ation but also the secondary heat flux (Figure 13.5, right graph). The simulated sec-ondary heat flux shows again significant differences when using the standard method with reduction factor compared to the detailed BSDF approach.

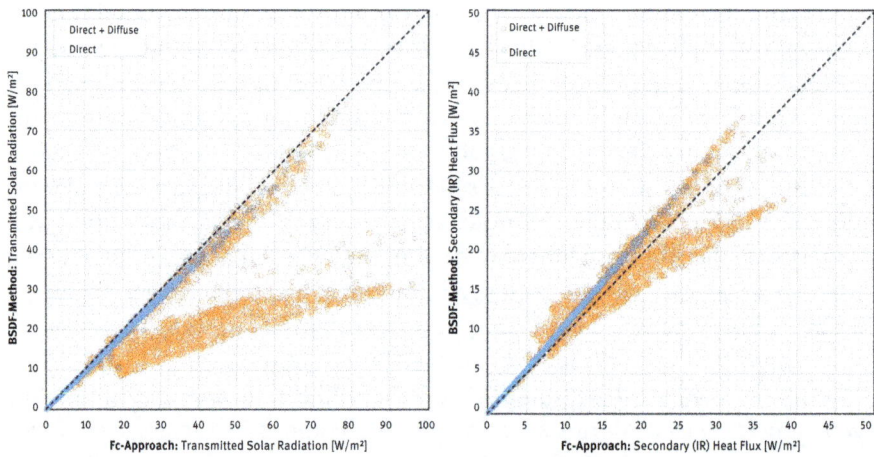

Fig. 13.5: Comparison of BSDF and Fc-method (Fc = 0.25) for modelling venetian blinds in BES Software, analyzing transmitted solar radiation (left graph) and secondary heat flux (right graph) (21). (modelled with TRNSYS 18)

13.3 Heat Transfer through Opaque Walls and Thermal Inertia Modelling

Accurately modelling the thermal mass of a building is essential for understanding its interactions with the ambient thermal environment. However, the complexity of buildings requires substantial simplifications to limit computation time for annual simulations. A fundamental simplification in building simulation is the *assumption of one-dimensional heat flux* through the envelope. This approach simplifies the modelling process by reducing heat flow through building envelopes to a singular, directional path perpendicular to the surface. While this makes the simulation process more manageable and computationally less demanding, it may not fully capture the complex thermal behaviours exhibited by real-world structures, particularly those with non-uniform material distributions and high thermal inertia.

Furthermore, the *methods used to model heat transport* through the building envelope require careful consideration. Common solution algorithms include the finite difference method and the conduction transfer function method. Each of these methods has its strengths and limitations and must be chosen based on the specific requirements of the building simulation.

13.3.1   Numerical Solution Techniques for Building Envelope Simulations

The Finite Difference Method (FDM) and Conduction Transfer Functions (CTF) are two basic numerical methods employed in building simulations to effectively handle the dynamic conditions of summer. Each method has its distinct characteristics and limitations that influence its application in various architectural and environmental scenarios.

FDM operates by approximating the heat equation on a discrete grid of nodes, which makes it particularly useful for situations where spatial variations in temperature need to be resolved over time. This method is adaptable enough to accommodate irregular materials and can adjust to various boundary conditions imposed by external weather conditions or indoor activities. For instance, FDM is capable of modeling Phase Change Materials (PCMs) by integrating their latent heat characteristics into the heat equation, although this requires a fine temporal and spatial resolution than usual to capture the phase change accurately. It is also effective in predicting temperature gradients within components, offering valuable insights into heat storage and release patterns crucial for managing peak loads.

On the other hand, the approach with **Conduction Transfer Functions** (CTF) simplifies the transient simulation of heat transfer through building components by transforming the time-domain heat conduction problem into the frequency domain using Laplace transformation. This approach is particularly well-suited for HVAC load calculations and assessing overall energy performance over extended periods. While CTFs are generally not suitable to handle non-linear behavior without significant modifications, they can estimate average temperatures within a component. However, they may not capture detailed temperature variations across the component accurately and only provide a macroscopic view of heat flows. Therefore, CTFs are better suited for systems where detailed spatial resolution is less critical. For applications requiring detailed modeling of, e.g., phase changes within a layer, or in scenarios involving complex thermal behaviors, FDM is recommended due to its ability to incorporate complex material properties and non-linear behaviors within a spatially resolved discretization. Conversely, for standard building simulations focusing on energy performance and HVAC loads, CTFs offer a computationally efficient alternative, albeit at the sacrifice of some detail in spatial and temporal resolution (22) .

In cases involving very thick building components, such as those found in historical buildings, neither FDM nor CTF may suffice alone. Here, a hybrid approach or a more sophisticated method like the Finite Element Method (FEM) might be necessary to accurately simulate the significant thermal mass and its impact on the building's thermal dynamics. These advanced methods enhance the ability of the simulation to capture complex thermal interactions more accurately, ensuring a more reliable assessment of building performance (22).

13.3.2 Challenges of One-Dimensional Modeling in Building Energy Simulations

In building simulation, the use of one-dimensional (1-D) heat transport simplification is required for computational efficiency. However, this approach can result in significant inaccuracies, especially in specific architectural contexts. For example, older buildings characterized by numerous thermal bridges typically experience substantial underestimations of heat loss or gain due to this simplification. Furthermore, buildings with complex geometric features such as curved walls or faceted façades present a challenge to the adequacy of 1-D modeling because heat transfer in these structures is not uniformly one-dimensional.

An illustrative example of the ***impact of dimensional considerations in the modeling of heat flow*** through the exterior walls is demonstrated with the case of a well-insulated manor house in Germany (23), depicted in Figure 12.6. Measurements from the building showed a level of room climate stability that could not be replicated in building simulation calculations using EnergyPlus. This discrepancy was primarily due to thermal bridges formed by deep window recesses and decorative elements similar to cornices and ceiling rosettes. These features were severely underestimated in the one-dimensional modeling, leading to a marked underestimation of the surface heat transfer and, consequently, the balancing heat flows. A subsequent analysis using a component simulation program, Delphin, clearly illustrated these discrepancies.

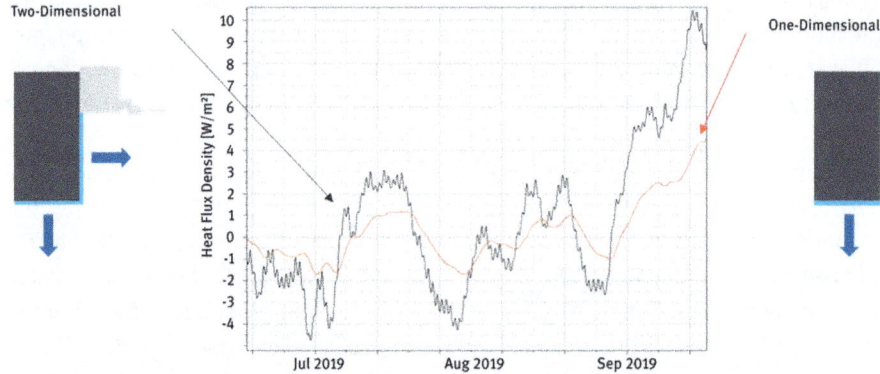

Fig. 13.6: Comparison of 2-Dimensional versus 1-Dimensional heat flow between the building component and room for a manor house in Tauchitz, Germany. (modeled with Delphin) (23,24)

The results as contained in Figure 13.6 visibly demonstrate that the underestimation of the heat exchange surface in the one-dimensional model (red curve in Figure 13.6) leads to a lower and faster heat exchange with the room compared to the two-dimensional view (black curve in Figure 13.6). This two-dimensional approach provides a more accurate representation of the actual heat flows and thermal interactions within the building, underscoring the limitations of simpler simulation methodologies for complex architectural structures.

Similarly, **buildings that incorporate active thermal components** like radiant cooling systems embedded within concrete slabs require detailed analysis. These systems entail complex interactions between the solid matrix and the embedded fluid flow, necessitating sophisticated multi-dimensional modeling to accurately predict their performance. Given these complexities, employing 2- or 3-dimensional modeling is often necessary to provide a more accurate assessment of building performance, particularly concerning thermal stability and energy efficiency to overcome the limitations of traditional 1-D simulation as used in BES, ensuring more reliable and comprehensive analysis and design solutions.

13.3.3 Modelling Latent Heat Storage (Phase Change Materials)

Incorporating **Phase Change Materials (PCMs)** into building envelopes can be an effective strategy for enhancing energy efficiency and thermal comfort within buildings. PCMs undergo a phase transition at a specific temperature, absorbing or releasing the latent heat which is specific to the material, therefore they can help to reduce indoor overheating (25).

PCMs are categorized based on their composition as organics such as paraffins and fatty acids, and inorganics typically comprising salt hydrates, and eutectics, which are combinations of materials. Organic PCMs are favored for their stable melting points and minimal supercooling, while inorganics offer higher latent heat capacities but can suffer from phase segregation and subcooling. Eutectics allow for tailor-made solutions with specific melting points (26,27).

In addition to their **specific latent heat of fusion,** the most important characteristic of PCM is their **melting and solidification temperature.** Commonly used organic PCMs have melting points ranging from 18°C to 28°C. The latent heat of fusion is typically around 200 kJ/kg for organics, significantly higher than their specific heat capacity in solid or liquid phases. In summer applications, the **effectiveness of PCMs** is influenced by their phase change temperature, ideally near the upper limit of the comfortable indoor temperature range to absorb excess heat efficiently. Effective night-time ventilation is crucial for the regeneration of PCMs, allowing them to solidify overnight and be ready for the next day's heat loads (28,29).

The challenge in the numerical **modeling of phase change materials** (PCMs) consists in identifying the depth at which phase change occurs. Therefor a **smaller than usual spatial discretization** is required. Typically, only the layers close to the boundaries are undergoing a change of phase, deeper layers are often not activated due to the change in material properties. If these effects are not represented in the simulation model, heat storage effects of PCM are massively overestimated. The most appropriate way of modeling phase transitions uses temperature-dependent functions for heat capacity and thermal conductivity during phase transition (25).

In addition to an appropriate spatial discretization when modelling PCM layers, the choice of timestep is also crucial for the accuracy of the simulation. Figure 13.7 displays the effect of different time steps (1 hour, 30 minutes, 15 minutes, and 5 minutes) on the temperature profile within a phase change material (PCM), simulated with esp-r (29). It is evident that a 1-hour timestep significantly distorts the temperature accuracy, proving to be inadequate for capturing the subtle dynamics of PCM behavior. As timestep size is reduced, the simulation results converge closer to expected outcomes, with minimal temperature discrepancies observed when transitioning from a 15-minute to a 5-minute timestep. The numerical model applied for the simulations shown in Figure 13.7 was previously developed and validated through extensive measurements in two real-size test chambers (30).

In summary, modeling of PCMs in buildings demonstrates that despite the complexity of phase change and associated thermodynamic processes, effective simulation is possible with appropriate numerical approaches and consideration of critical parameters. The challenge lies in refining models sufficiently to make realistic predictions while maintaining computational efficiency.

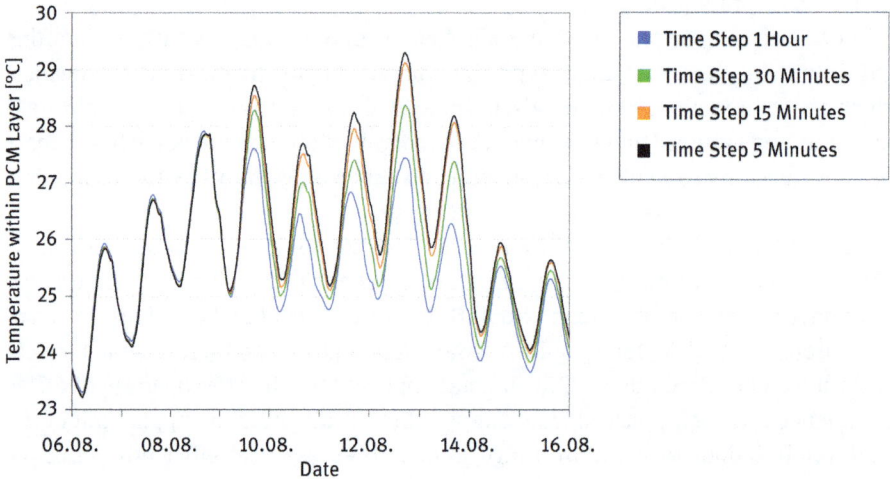

Fig. 13.7: Impact of timestep on calculated Phase Change Material (PCM) temperature according to (29). (modelled with esp-r)

13.4 Natural Ventilation Modelling

In buildings where there is no mechanical ventilation available, the manual operation of windows provides the required ventilation rates. In addition to ensuring enough fresh air for occupants, ventilation has an important impact on the thermal conditions in rooms during summertime. Particularly at times when outdoor temperatures are below indoor temperatures which is often the case during nighttime, high ventilation rates help to cool down the building. Hence the modeling of natural ventilation through windows is crucial for accurately predicting energy consumption and thermal comfort during periods of hot outdoor conditions.

While mechanical ventilation systems are operated with specified volume flows, natural ventilation is caused by stack effect and/or by wind pressure. Natural ventilation of indoor spaces is typically provided through operable windows. Air change rates in summertime can be increased through cross ventilation by opening doors or windows on more than one side of the rooms. Cross ventilation is usually dominated by wind pressure unless wind speed is very low (31). Air flow through vertical vents such as solar chimneys or ventilated facades however are predominantly driven by stack effect. There are two ways of *modelling natural ventilation*:

1. The *simplified approach* assumes a user-specified air exchange rate without considering if this air exchange can happen under given conditions. The advantage of this approach is that temperature-dependent control can be applied with little effort.
2. An *air flow network* considers pressure differences (temperature and/or wind induced) between zones. Air flow is calculated between nodes and through components which are represented through pressure loss. In general air flow networks require a good understanding of the properties of inlet and outlet components.

To demonstrate the importance of different natural ventilation approaches on the overheating assessment results, a room in a residential building was simulated in IDA ICE applying different air exchange models. The following natural ventilation profiles are compared:

a) *Constant ventilation rate* (neglecting pressure-driven and temperature-driven air exchanges) according to standard DIN 4108-2 (air exchange rate of 0.5 h^{-1} as standard, if room temperature is above 23 °C and above outdoor air temperature an air exchange rate of 3,0 h^{-1} can be applied from 6 am to 11 pm and 2,0 h^{-1} from 11 pm to 6 am)
b) *Wind- and temperature-driven air exchanges* according to the airflow network model with the assumption that:
    i. *one window* is *fully opened for the whole day* (if room temperature is above 23 °C and above outdoor air temperature)

ii.   *one window* is *fully opened from 6 pm to 6 am* (if room temperature is above 23 °C and above outdoor air temperature) representing that the flat is not occupied during the day

iii.   *both windows* are only *tilted from 6 pm to 6 am* (if room temperature is above 23 °C and above outdoor air temperature) representing a realistic window ventilation behavior (32)

The ventilation rates and evolving operative room temperatures of the different approaches are depicted in Figure 13.8.

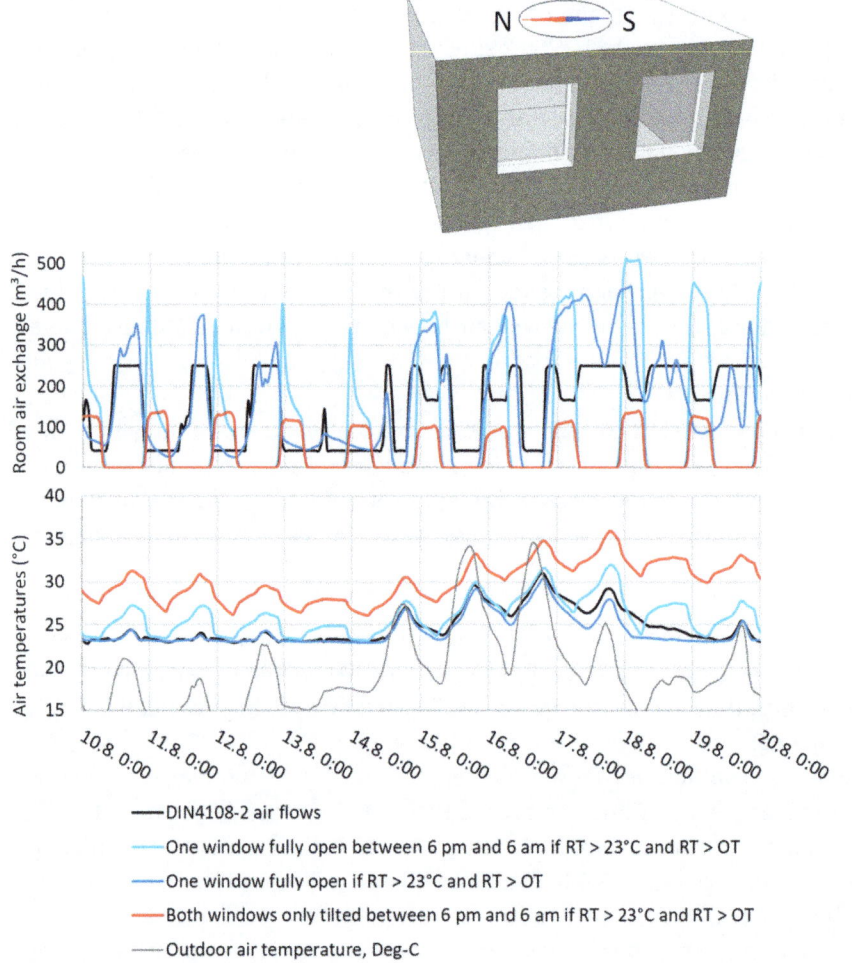

Fig. 13.8: Impact of different natural ventilation modelling on air exchange rate and room temperature for the depicted room (27 m² large with two 2 m² large windows) for 10 days in summer (simulated with IDA ICE).

A first significant difference between the approach a) and b) is that air exchange rates are not constant if an airflow network model is applied reflecting that the air exchange is dependent on the temperature differences between indoor and outdoor as well as on the wind conditions. Second, the air exchange rate is comparable between a) constant air exchange rates and b)-i the airflow network model if one window is fully open the whole day, but very different for b)-ii when the windows are only opened from 6 pm to 6 am. If the windows are only tilted as in b)-iii much lower ventilation rates are obtained. These differences in air exchange have a strong impact on the evolving indoor room temperatures ranging in maximum from 31 °C to 36 °C. This simple example highlights that simplified assumptions of constant air exchange rates (neglecting pressure-driven and temperature-driven air exchanges) are insufficient for a realistic overheating assessment. This is particular true if one considers that opening the window the whole day is not a representative, realistic window ventilation behavior (32).

That the used airflow network model represents air exchange by window ventilation in a realistic manner was verified by comparing BES results with indoor temperature monitoring of a multi-residential building (33). As a result, a good match between the measured and simulated room temperatures was achieved for an entire summer for different rooms of the multi-residential building (33).

14 Conclusion

In conclusion, this chapter has elucidated the integral ***role of advanced simulation models in assessing and mitigating overheating risks in buildings***, particularly under the looming impact of global warming. By delving into the specifics of various simulation tools and methods, such as EnergyPlus and IDA ICE, the discussions have underscored the necessity of incorporating detailed, dynamic models that can navigate the complexities of thermal interactions and environmental impacts on buildings. The ***exploration of different mitigation strategies***, including the use of ***phase change materials***, enhanced ***ventilation schemes***, and ***innovative glazing*** and ***shade technologies***, highlights the progressive approaches towards enhancing building resilience against overheating. These strategies, backed by rigorous simulation and validation, present promising avenues for architects and engineers to design buildings that are not only energy-efficient but also comfortable and sustainable in the face of changing climate conditions. Moreover, the ***critical analysis of the simulation tools' capabilities*** in reflecting true building dynamics offers valuable insights into the ongoing enhancements needed in simulation technologies. This ensures that they stay relevant and effective in predicting building performance accurately, aiding in the optimization of design and operational strategies to combat overheating risks.

Ultimately, as this chapter suggests, the ***continual advancement in simulation methodologies and the integration of empirical data for validation are essential***. This approach will bridge the gap between theoretical predictions and actual building performance, enabling the construction industry to respond more adeptly to the environmental challenges of the future. The collaborative efforts between researchers, tool developers, and practitioners in refining these simulation tools will undoubtedly play a pivotal role in shaping the next generation of energy-efficient and climate-resilient buildings.

15 References

1. ENVI-met. High-Resolution 3D Microclimate Modeling for Climate Adaptation: ENVI-met GmbH [cited 2024 Jun 19]. Available from: URL: https://envi-met.com/.
2. EN ISO 13791: Thermal Performance of Buildings - Calculation of Internal Temperatures of a Room in Summer Without Mechanical Cooling - General Criteria and Validation Procedures: International Organization for Standardization; 2012 2012.
3. VDI 6007 Blatt 1. Berechnung des instationären thermischen Verhaltens von Räumen und Gebäuden - Raummodell; 2015 [cited 2024 Apr 28]. Available from: URL: https://www.vdi.de/richtlinien/details/vdi-6007-blatt-1-berechnung-des-instationaeren-thermischen-verhaltens-von-raeumen-und-gebaeuden-raummodell-1/.
4. ASHRAE Standard 140-2017. Standard Method of Test for the Evaluation of Building Energy Analysis Computer Programs. Atlanta, GA: Refrigerating and Air-Conditioning Engineers; 2017 2017.
5. VDI 6020. Anforderungen an Rechenverfahren zur Gebäude- und Anlagensimulation. Berlin: VDI-Verlag; 2001 2001.
6. Hirth S. SimQuality: A Novel Test Suite for Dynamic Building Energy Simulation Tools. In: Proceedings of the IBPSA Building Simulation Conference; 2021 [cited 2021 Aug 14]. Available from: URL: https://cloudstore.zih.tu-dres-den.de/in-dex.php/apps/files/?dir=/Shared/Publikationen/2021_BuildingSimula-tion&fileid=412661053#pdfviewer.
7. Weiß D, Hirth S, Nicolai A, Nouri A, Agudelo J, Rolffs R. SimQuality; 2020. Available from: URL: https://simquality.e3d.rwth-aachen.de/de/.
8. Sim Quality Interactive Testsuite: SIM-Quality Dashboard; 2022 [cited 2024 Apr 28]. Available from: URL: https://simquality-dashboard.onrender.com/.
9. Ruiz GR, Bandera CF. Validation of Calibrated Energy Models: Common Errors. Energies 2017; 10(10):1587.
10. Freudenberg P, Dresel MD, Batayneh Z. Assessing Statistical Indices for Calibrating Building Performance Simulation Models via Continuous Time Series. J. Phys.: Conf. Ser. 2023; 2654(1):12056. Available from: https://doi.org/10.1088/1742-6596/2654/1/012056.
11. U.S. Department of Energy. EnergyPlus: Whole Building Energy Simulation Program; 2022 [cited 2024 Jun 19]. Available from: URL: https://energyplus.net.
12. Thermal Energy System Specialists, LLC. TRNSYS: Transient System Simulation Tool [cited 2024 Jun 19]. Available from: URL: http://www.trnsys.com.
13. ESRU, University of Strathclyde. ESP-r: A Building Energy Simulation Tool [cited 2024 Jun 19]. Available from: URL: https://www.esru.strath.ac.uk/Programs/ESP-r.htm.
14. EQUA Simulation AB. IDA Indoor Climate and Energy (IDA ICE) [cited 2024 Jun 19]. Available from: URL: https://www.equa.se/en/ida-ice.
15. Clarke JA. Energy Simulation in Building Design. Revised Edition. London: Routledge; 2015.
16. Hensen JL, Lamberts R. Building Performance Simulation for Design and Operation. 2nd ed. Abingdon, Oxon, New York, NY: Taylor & Francis; 2019.
17. Ganji Kheybari A, Hoffmann S, Lee S-Y. Simulation-Based Framework Exploring the Controls for a Dynamic Facade with Electrochromic Glazing (EC). In: PowerSkin 2019 Proceedings; 2019.
18. Ganji Kheybari A, Steiner T, Liu S, Hoffmann S. Controlling Switchable Electrochromic Glazing for Energy Savings, Visual Comfort and Thermal Comfort: A Model Predictive Control. CivilEng 2021; 2(4):1019–53.

19.   Lee ES, Pang X, Hoffmann S, Goudey H, Thanachareonkit A. An empirical study of a full-scale polymer thermochromic window and its implications on material science development objectives. Solar Energy Materials and Solar Cells 2013; 116:14–26.

20.   Hoffmann S, Lee ES, Clavero C. Examination of the technical potential of near-infrared switching thermochromic windows for commercial building applications. Solar Energy Materials and Solar Cells 2014; 123:65–80.

21.   Hoffmann S, Kheybari AG. Untersuchungen zum sommerlichen Wärmeschutz – Teil 2: Vergleich zwischen Modellierung mit Abminderungsfaktor (F C -Faktor) und bidirektionalem Ansatz (BSDF-Methode). Bauphysik 2021; 43(2):87–99.

22.   Pilet T, Rakha T. Modeling of transient conduction in building envelope assemblies: A review. Science and Technology for the Built Environment 2022; 28(6):706–16.

23.   Technische Universität Dresden, Institut für Bauklimatik. Erhalt und Bewirtschaftung von Burgen und Schlössern in Mitteldeutschland Angesichts Sich Verändernder Energetischer und Klimatischer Randbedingungen: Schlussbericht. Dresden; 2021.

24.   Grunewald J, Nicolai A, Paepcke A, Fechner H. 'DELPHIN.' Dresden: Bauklimatik-Dresden; 2023 [cited 2023 Aug 2]. Available from: URL: : https://www.bauklimatik-dresden.de/delphin/.

25.   Hoffmann S. Numerische und Experimentelle Untersuchung von Phasenübergangsmaterialien zur Reduktion Hoher Sommerlicher Raumtemperaturen; 2006 [cited 2024 May 1]. Available from: URL: https://e-pub.uni-weimar.de/opus4/frontdoor/index/index/year/2007/docId/823.

26.   Al-Yasiri Q, Szabó M. Incorporation of phase change materials into building envelope for thermal comfort and energy saving: A comprehensive analysis. Journal of Building Engineering 2021; 36:102122.

27.   Rida M, Hoffmann S. The influence of macro-encapsulated PCM panel's geometry on heat transfer in a ceiling application. Advances in Building Energy Research 2022; 16(4):445–65.

28.   Mazzeo D, Romagnoni P, Matera N, Oliveti G, Cornaro C, Santoli L de. Accuracy Of The Most Popular Building Performance Simulation Tools: Experimental Comparison For A Conventional And A PCM-Based Test Box. In: Proceedings of the IBPSA Conference; 2019 [cited 2021 Jun 11]. Available from: URL: http://www.ibpsaa.org/proceedings/BS2019/BS2019_210381.pdf.

29.   Hoffmann S, Kornadt O. An Investigation on Phase Change Materials to Reduce Summer Overheating. In: Proceedings of the IBPSA Canada Esim Conference; 2006.

30.   Hoffmann S, Kornadt O. Simulation of Phase Change Materials - Validation of the Numerical Model. In: Proceedings of the First German-Austrian IBPSA Conference; 2006.

31.   Schmidt D, Maas A, Hauser G. Experimental and Theoretical Study on Cross Ventilation - Designing a Mathematical Model. Nordic Journal of Building Physics.

32.   Schünemann C, Schiela D, Ortlepp R. How window ventilation behaviour affects the heat resilience in multi-residential buildings. Building and Environment 2021; 202:107987.

33.   Schünemann C, Schiela D, Ortlepp R. Guidelines to Calibrate a Multi-Residential Building Simulation Model Addressing Overheating Evaluation and Residents' Influence. Buildings 2021; 11(6):242.

Christoph Schünemann, Peggy Freudenberg, Tim Felix Kriesten
Indoor Overheating Assessment

Analyzing and Refining Indicators for Indoor Overheating: From Existing Standards to a Combined Approach

Abstract In the field of indoor overheating analysis, the development of valid indicators and thresholds is recognized as crucial for assessing overheating risk or heat resilience in rooms and buildings. This assessment process is typically undertaken through methods such as building performance simulation or indoor monitoring. This chapter focuses on exploring the existing criteria and thresholds derived from a variety of standards and methodologies, which are pertinent to the assessment of overheating in free-running buildings. The goal is to establish a refined set of indicators for the assessment of residential buildings, integrating the strengths of existing methodologies. In the proposed indicator set, several key elements are integrated: temperature exceedance hours, weighted for their significance (as informed by ASHRAE 55 and DIN 4108-2), adaptive temperature thresholds that reflect the adaptive nature of human thermal sensation (based on EN 16798-1 and ASHRAE 55), and a criterion for night-time indoor temperatures (sourced from CIBSE TM52 and TM59). The variability in room usage, as outlined in CIBSE TM59, is also considered and extended. The effectiveness of this indicator set was evaluated through simulations of an apartment building in various summer regions of Germany. These results were then compared with the outcomes derived from existing overheating assessment standards, specifically CIBSE TM59, CIBSE Guide A, and DIN 4108-2. The chapter concludes by discussing the complexities involved in formulating more valid and reliable thresholds for overheating assessment. Future enhancements in this area should include the integration of insights from public health literature and the consideration of indicators considering humid heat sensation.

Keywords: Indoor Overheating Indicators, Indoor Overheating Assessment, Thermal Comfort, Indoor Climate, Building Performance Simulation, Heat Stress Analysis

Christoph Schünemann, Leibnitz Institute of Ecological Urban and Regional Development (IOER), Weberplatz 1, 01217 Dresden, +49(0)351 4679-194, c.schuenemann@ioer.de
Peggy Freudenberg, Dresden University of Technology, Institute of Building Climatology, Zellescher Weg 17, 01062 Dresden, +49(0)351 463-35259, peggy.freudenberg@tu-dresden.de
Tim Felix Kriesten, Leibniz Institute of Ecological Urban and Regional Development (IOER), Weberplatz 1, 01217 Dresden, +49 (0)351 4679-194, t.kriesten@ioer.de

16 Introduction

In the evolving field of building performance analysis, the assessment of overheating risk stands out as a critical area of study, particularly as climate variability intensifies and indoor comfort standards become more stringent. The ability to accurately predict and mitigate overheating in residential and commercial buildings not only enhances occupant comfort but also contributes to energy efficiency and sustainability goals. This chapter is dedicated to exploring the sophisticated methodologies used in setting thresholds for the evaluation of overheating risks based on simulation and measurement data, which primarily focus on the hourly trajectories of *operative temperature*.

The foundation of this discussion lies in understanding the nuanced approaches to *defining threshold values* that can reliably indicate the risk of overheating. These thresholds are pivotal in building simulations and real-time monitoring, serving as benchmarks against which the adequacy of a building's thermal environment can be measured. The chapter delves into the conceptual frameworks that underpin both *absolute and model-based thresholds* derived from adaptive and static approaches. Each method offers unique insights and operational advantages, depending on the specific building context and climatic conditions. Moreover, the *criteria for exceeding these thresholds* are examined in detail. The design of these criteria can range from stringent to flexible, reflecting the diverse needs of building users and the variability of environmental conditions. Such designs are crucial for accommodating the daily and seasonal fluctuations that characterize the indoor climate. This chapter also discusses the *temporal aspects of threshold evaluation*, which include determining whether assessments should cover the entire year, focus on the peak summer period, or target specific times of day or night. This temporal differentiation is essential for tailoring the assessment to the most relevant and critical periods of potential overheating. Additionally, the chapter considers the *user-centric aspects* of threshold setting, acknowledging that different occupants may have *varying requirements* based on their *usage patterns* and personal preferences. Integrating these diverse user needs into the threshold-setting process enhances the relevance and applicability of the overheating assessments (OA).

Through a comprehensive analysis, this chapter aims to provide a deeper understanding of the complex interplay between building performance, environmental conditions, and human factors in the assessment of overheating risks. It sets the stage for *introducing a refined approach to threshold setting* that incorporates the best practices from existing standards and adapts them to the nuanced needs of different environments and user groups. This approach not only aims to improve the accuracy of overheating risk assessments but also to enrich the strategies for managing and mitigating these risks in diverse building settings.

17 Operative Temperature in Comfort Assessment

The first chapter of this book highlights the limitations of using static respectively heat-balance-based approaches for indoor comfort assessment, particularly during extreme summer seasons. The empirical approaches have proven to be more suitable here, as there is a correlation between the outside air temperature and the acceptable room temperature for all building types. Integrating this understanding into practical applications, it is often the indoor air temperature that is used to gain information about heat stress and overheating intensity in rooms and buildings. This is mainly due to the ease of measuring air temperature. However, this approach is limited as it does not consider the full spectrum of physical properties.

This brings into focus the need for **empirical indoor comfort assessment approaches** that **rely on operative temperature** instead of just air temperature. The operative temperature is a more comprehensive indicator as it considers both the convective and long-wave radiative heat exchanges between the body surface and the environment. This is especially crucial in conditions where the surface temperatures of surrounding structures significantly differ from the air temperature. High radiant temperatures are common for surfaces exposed to solar irradiation, especially when combined with high surface absorption. Conversely, a space can feel cooler if the air is ventilated through windows, yet the building components retain warmth due to their heat storage capacity. Given these factors, the operative temperature, θ_{op}, serves as a **central indicator for overheating assessment (OA)** in international standards. It **combines** both **air temperature,** θ_{air}, and **radiant temperature,** θ_{rad}, into a single variable. Depending on the selected level of detail, the operative temperature can be determined in different ways, e.g. weighted via the radiative h_{rad} and convective h_{conv} surface transfer coefficients (variant 1), as an arithmetic average (variant 2) or via the air velocity v_{air} as an input parameter accounting for the convective transfer (variant 3) in accordance with (1–3):

$$\theta_{op} \cong \frac{h_{rad} \cdot \theta_{rad} + h_{conv} \cdot \theta_{air}}{h_{rad} + h_{conv}} \cong \frac{\theta_{rad} + \theta_{air}}{2} \cong \frac{\theta_{rad} + \sqrt{10 \cdot v_{air}} \cdot \theta_{air}}{1 + v_{air}}$$

The variability of indoor air velocities across Europe is huge, especially during the summer months in non-air-conditioned buildings with or without fans. In general, indoor **air velocities** in European countries are assumed to range between 0.15 to 0.40 meters per second (m/s) in summer (4). For the resulting operative temperature from the above approach, the air temperature multiplier rises to a weighting factor of more than 1.7. For buildings without air conditioning, the use of passive cooling strategies like ceiling or desk fans is common. They improve summer thermal comfort as the airflow speed and direction can often be manually controlled by the occupants. Resulting airflow speeds in the room depend strongly on the distance between the fan and the occupants as well as on the fan rotating speed. Resulting velocities achieve values of up to 0.5 m/s and in some cases up to 1.8 m/s, which represents weighting factors of up to 2.2 and exceptionally up to 4.2 for the above-mentioned approach for the operative temperature (5).

It should also be recognized that the influence of the **radiation temperature** depends heavily on the room characteristics and the user's location in the room. For example, the interior surface temperatures of façade glazing can be up to 20 Kelvin higher than the air temperature during the summer (6). In such cases, the influence of the radiation temperature increases with increasing closeness to the façade and an arithmetic between air and radiant temperature is no longer representative. In addition, for the detailed calculation of mean radiant temperature (θ_{rad}), a method involving the direct use of radiant intensities at a specific point in a room is considered more accurate than the use of surface emissions resp. temperatures with view factors. This approach is based on the localized radiant intensity (7).

However, it's important to consider that while the **operative temperature** offers a more holistic view, it **simplifies some aspects**. For instance, it does not account for the impact of **humidity** on thermal comfort. Humidity plays a significant role in comfort levels during summer, as it affects sweat production and evaporation from the skin. These factors are crucial in hot and humid conditions, which is typically not the case in Europe, but are not directly considered in the operative temperature calculation. It must also be noted that the operative temperature as the sole characteristic value of the room climate does not allow any consideration of **personal characteristics**, for example clothing insulation and activity level (metabolic heat generation), and thus only assesses the thermal comfort properties of a room neglecting adaptive human behavior or differences in perception.

Operative temperature is used in various national and international standards **for assessing building overheating risk** and year-round thermal comfort. For overheating risk, the Chartered Institution of Building Services Engineers (CIBSE) in the UK has developed **standards** such as TM52, which specifies limits for operative temperature during summer in general, and TM59 for the specifics of residential buildings (8,9). The German DIN 4108-2 standard also addresses overheating risk, considering the operative temperature in its assessment criteria (10). The same applies to the Swiss standard SIA 180 and the Austrian standard ÖNORM B 8110-3 (11,12).

Regarding **year-round thermal indoor comfort**, standards include furthermore the American Society of Heating, Refrigerating, and Air Conditioning Engineers (ASHRAE) Standard 55, CIBSE guidelines, and the European EN 16798-1:2019, which replaced EN 15251. These standards utilize operative temperature as a key factor in evaluating the indoor thermal environment, ensuring comfort and energy efficiency in building design.

18 Threshold Specification

In principle, different approaches can be considered for the definition of limit values for the overheating risk assessment in summer based on simulated or measured time series of the operative temperature. The limit value can be defined as a ***time-constant limit value*** or as a varying ***model-based limit value*** (e.g. from a heat balance-based model like Fangers PMV or an empirical model like in EN 16798-1, former 15251). It is also possible to consider different ***time slots*** (e.g. night, day) for the assessment of fulfillment of the limit value (e.g. presence of persons in the considered room). Additionally, there is the question of the ***specified deviation tolerance*** (e.g. strict limit value, relative deviation).

    Time-constant temperature limits, such as a maximum indoor temperature of 30°C, provide a clear, straightforward benchmark. This approach is particularly effective in environments where variability in user preferences and indoor climate conditions is minimal. For physically demanding working conditions, for example, where the focus is on maintaining conditions that are compatible with health, absolute limit values ensure a safe working environment. The approach of time-constant limits is therefore especially useful to define *tolerability* limits. In contrast, ***model-based approaches*** consider a variety of depending factors. In the case of heat balance-based approaches like Fanger PMV, they include individual user properties (like metabolic rate and clothing) and indoor climate conditions (such as humidity and air movement). In the case of empiric approaches, they include indirectly the interaction between the occupant and the building as well as occupant properties, both aligned with the ambient climate conditions. These models are adaptable. The Fanger PMV model, for example, calculates the predicted mean vote of thermal comfort based on six parameters, offering a dynamic and user-centric approach. Similarly, standards like EN 16798-1 utilize empirical models that factor in regional climatic variations and adaptive comfort principles, recognizing that comfort perception varies based on local climate and perceived past climate conditions.

    The key advantage of model-based approaches is their flexibility and adaptability to a wide range of scenarios and individual preferences. They are particularly useful in diverse settings like offices or residential buildings, where occupant comfort is strongly time-depending, subjective, and influenced by multiple factors.

    In addition, ***model-based approaches*** can also be applied to ***derive time-constant limit values*** for a selected user group and assumed time-constant indoor or outdoor climate conditions. This approach is followed by some standards in the application of the Fanger PMV model, which is then reduced to seasonal recommended values of the operative temperature e.g. in ASHRAE 55 or ISO 7730. More complex heat-balanced approaches can be simplified in the same manner and thus made easy to apply. However, it should be noted that the applicability for free-running buildings with fluctuating indoor climate conditions and variable individual characteristics of the users is no longer expedient. On the other hand, the simplicity of absolute thresholds is advantageous for broad applications where specific health-related tolerability limits are required, or where the indoor climate and user interaction are consistent and predictable. In such cases, fixed limit values offer a practical solution to ensure a basic level of tolerability or comfort while maintaining simplicity in application.

An insight into the adaption of these time-constant or model-based limit values by diverse national codes is given in Table 18.1.

Tab. 18.1: Overview of time-resolved limit types for indoor operative temperature in selected overheating assessment standards

Limit type	National Code	Specification
Time-constant limit	DIN 4108-2:2013	Depending on the location 25°C, 26°C or 27°C for all buildings
	ASHRAE 55: 2023	23°C to 26°C based on user properties like clothing insulation and activity (derived from Fanger PMV) for buildings with an air conditioning system
	CIBSE TM59: 2017	26°C for residential buildings
	CIBSE Guide A:2006	25°C to 28°C for comfort and 30°C as tolerability limit for offices resp. working conditions
Model-based limit	CIBSE TM52: 2013 (Criterion 1 to 3)	Adaptive model-based limit value (see ISO 74: 2014 resp. EN 15251:2007) depending on past outdoor air temperature
	SIA 180: 2014	Adaptive model-based limit value (see ISO 74: 2014 resp. EN 15251:2007) depending on past outdoor air temperature
	ASHRAE 55: 2023	Adaptive model-based limit value depending on past outdoor air temperature

Besides the characteristic of the threshold value, the **tolerance for deviations** also plays a role in defining limits for indoor climate. An insight into the tolerance limits of selected standards is provided in Table 18.2.

Setting **strict, absolute limits** that must not be exceeded at any time during the observation period ensures minimum standards. The advantage of these limits is their clear and unambiguous specification, which facilitates compliance and verification. However, their disadvantage lies in the lack of flexibility, as they **do not allow for temporary or minor exceedances**, which is often unrealistic in practice. To address this issue, it is sensible to combine these inflexible limits with empirical models that yield higher thresholds for the summer season as implemented in SIA 180. This approach can create a certain robustness against individual outliers. This integration of strict absolute limits with more dynamic, empirically informed thresholds offers a more nuanced and realistic framework for managing and evaluating indoor climate parameters.

The use of **relative limits** offers more flexibility in this respect. These can be defined **in relation to the evaluation time,** for example, as a percentage of the time intervals in which the criteria are exceeded. This allows for better adaptation to varying environmental conditions and can provide more realistic results. However, a disadvantage might be a certain vagueness of the criteria and the potential neglect of short-term but significant deviations. Relative thresholds are for example part of the CIBSE TM59 and TM52.

Alternatively, limits can also be expressed as an **integral value,** where the product of the exceedance time and amount is calculated in Kelvin-hours. This method considers the duration and extent of exceedances and provides a comprehensive assessment of the indoor climate. However, a disadvantage might be the more complex calculation and interpretation, which requires specific expertise. Integral tolerances are defined in the DIN 4108-2 and CIBSE TM52.

Tab. 18.2: Overview of tolerance limit types for indoor operative temperature in selected overheating assessment standards

Limit type	National Code	Specification
Absolute	CIBSE TM52: 2013 (Criterion 3)	Exceedance of the model-based (see ISO 74: 2014 resp. EN 15251:2007) limit value by less than **4 K** throughout the year
	SIA 180: 2014	Exceedance of the model-based (see ISO 74: 2014 resp. EN 15251:2007) limit value by less than **3 K** throughout the year
Relative	CIBSE Guide A:2006	Exceedance of the time-constant limit value of 28 °C by **less than 1%** per annual period during occupied hours
	CIBSE TM52: 2013 & CIBSE TM59: 2017 (both Criterion 1)	Exceedance of the model-based (see ISO 74: 2014 resp. EN 15251:2007) limit value by more than 3 K at **less than 3%** during occupied hours in summer (May to September)
	CIBSE TM59: 2017 (Criterion 2)	Exceedance of the time-constant limit value of 26°C for less than **1%** during the night (10 pm to 7 am) in bedrooms in summer (May to September)
Integral	DIN 4108-2:2013	Exceedance of the time-constant limit value by less than **500 Kh** for residential buildings and **1200 Kh** for all other buildings for the entire year
	CIBSE TM52: 2013 (Criterion 2)	Exceedance of the model-based (see ISO 74: 2014 resp. EN 15251:2007) limit value by less than **6 Kh** in any one day of the year for non-residential buildings

Since in practice the question often arises as to how representative the various comparative parameters are and to what extent they agree with each other, the evaluation parameters were compared below in Figure 18.1 using an open dataset (13), which includes indoor climate measurements in eleven naturally ventilated offices over a period of three years and a time resolution of 10 minutes. A key finding of this comparison is the fact that both the relative and integral exceedance of the selected time-constant limit value of 26°C correspond poorly with the measurements of the maximum air temperature values. Extreme values are therefore a weak indicator for assessing the risk of overheating based on measured values. This correlation could be different in simulation results because consistent user behaviors are used in these. Furthermore, the comparison shows that relative and integral exceedance are very similar, and therefore both parameters can be used equivalently to assess the risk of overheating.

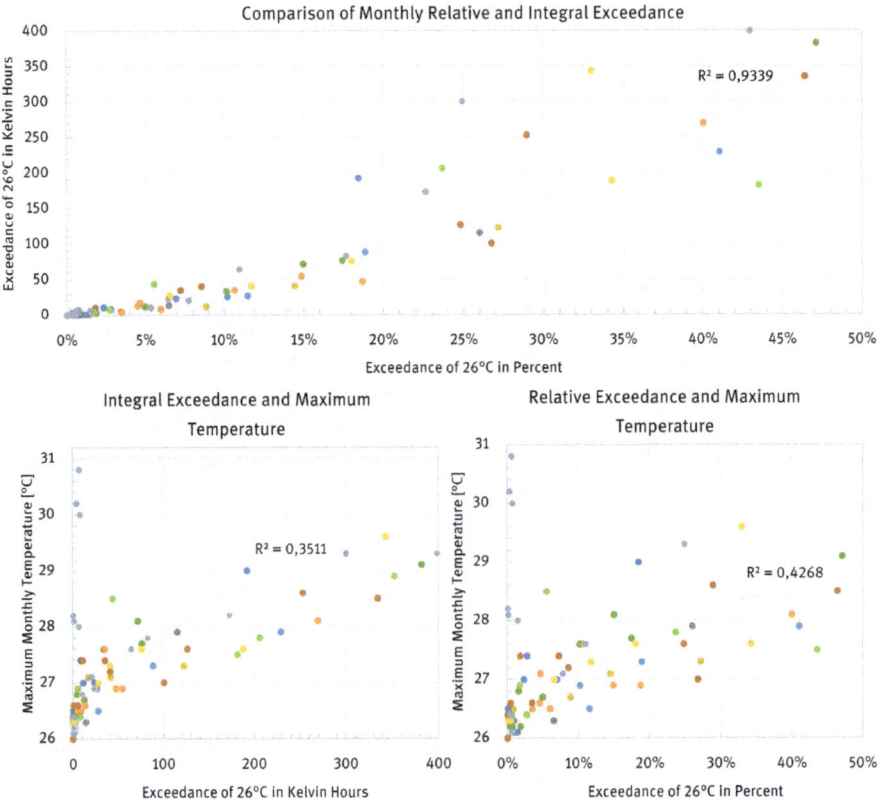

Fig. 18.1: Comparison of exceedance indicators for an open-access dataset that covers three years of indoor air temperature measurements in eleven office rooms. The top graph shows the correspondence between relative and integral exceedance, the bottom graph left side shows integral and absolute values, bottom right side shows relative and absolute value. The polynomial trend line of all scatter points is given in the background with its coefficient of determination (R^2).

However, it should be noted that the ***time resolution*** of the measurement or simulation time series can be particularly important for the extreme values. The lower the selected temporal resolution the more extreme the individual values can occur. This relation is low in the analyzed data set. If the comparison is carried out for the hourly averaged temperature values, the coefficient of determination increases only slightly, by approx. 0.01 due to the smoothing effect of hourly averaging. Surprisingly, the relationship between the relative and integral frequency distributions is reversed with decreasing temporal resolution. Here, the coefficient of determination decreases by about 0.02 with increasing time steps. For more general findings on these correlations, analyses of numerous data sets would be required, including in different building types. These analyses would also need to consider the influences of the occupancy time.

The final aspect of data analysis involves *selecting the evaluation time window*. Standards offer various approaches, ranging from year-round temperature profile evaluations to assessments of summer phases and even specific summer days or nights. Additionally, evaluations can be limited to the occupancy time within the evaluation window. This selection critically influences how representative and meaningful the overheating risk assessment results are. Year-round evaluations capture seasonal differences in a building's thermal performance, revealing its ability to cope with varying climatic conditions throughout the year. Focusing on summer phases or specific summer days, however, allows for a more detailed analysis of critical conditions. Limiting evaluations to typical summer days or nights is particularly relevant in regions with extreme temperature fluctuations between day and night. In such cases, it's vital to consider both peak loads during the day and the building's cooling potential at night. These dual considerations are crucial for developing strategies to improve thermal comfort and reduce overheating risk. Moreover, considering real occupancy times in the evaluation window is essential. In buildings primarily used during the day, such as offices or schools, it's important to analyze temperature profiles during working or school hours. In contrast, evening and night hours are often more relevant in residential buildings, as these represent the main usage times.

In *conclusion*, approaches to assess summer overheating risk involve a nuanced balance between different types of *threshold limits, adaptability to various scenarios, and thorough consideration of user preferences and occupancy times*. Time-constant limits provide straightforward benchmarks in environments with minimal variability, while model-based approaches offer flexibility and adaptability, accounting for individual user characteristics or outdoor climate conditions. The decision between adopting time-constant or model-based limits varies according to national standards, each with its own set of criteria. Absolute limits, though clear and specific, lack flexibility, and therefore, combining them with empirical models creates a more robust framework. This combination effectively manages both strict health-related tolerability and subjective comfort requirements. Relative limits, defined by the percentage of time intervals exceeded, and integral values, which consider the duration and extent of exceedances, provide additional layers of flexibility and comprehensiveness. However, they require careful consideration to avoid overlooking short-term but significant deviations.

19 Proposal for a Refined Indicator Set in Residential Buildings

In this chapter, a comprehensive approach is proposed for the assessment of overheating risks and the maintenance of thermal comfort within residential buildings, with an emphasis on sleeping areas. This approach integrates a variety of existing assessment models and thresholds, which are tailored to meet the unique needs of residential environments. The indicators and thresholds from CIBSE TM52, CIBSE TM59, DIN 4108-2, and EN 16798 are combined in this refined approach.

The Model-Based **Adaptive Indoor Threshold,** grounded in the empirical indoor operative temperature threshold model as outlined in EN 16798, is adopted. This model utilizes comfort categories based on ASHRAE 55, similar to those used in CIBSE TM59, and considers multiple factors including local climatic conditions and specific user characteristics. It offers flexibility for adaptation to various residential scenarios, particularly suitable for the variable usage conditions in residential buildings. The acknowledgment that human heat sensation varies with past weather conditions contrasts with the constant indoor operative temperature threshold assumption seen in standards like DIN 4108-2. The adaptive comfort model already indirectly considers different temperature limits for different climate zones with different summer conditions, thus making the specification of different indoor temperature limit values for different summer regions unnecessary, as is done in DIN 4108-2. Unlike in CIBSE TM59, it is suggested that category I in EN 16798 (high comfort) is used instead of category II (standard comfort) for the adaptive comfort model, due to the high indoor temperature limits that category II allows.

The employment of **a combination of an integral threshold** (measured in Kelvin-hours for comfort assessment) **and an absolute threshold** (for defining tolerability limits) is proposed. The integration of temperature-weighted exceedance hours (TWEH), which consider both the frequency and the intensity of indoor temperature exceedance above a given threshold, is suggested. The integral threshold allows for a comprehensive evaluation of thermal comfort over extended periods, while the absolute threshold acts as a safety net to mitigate health risks associated with extreme temperature conditions.

Room-specific usage characteristics are considered, motivated by the distinct usage times of different spaces in residential buildings, inspired by the methodology in CIBSE TM59. This room-wise consideration enables a more accurate and realistic assessment of overheating risk and thermal comfort. Occupancy times are categorized into two types: one with a usage duration of 10 hours per day for living rooms and kitchens, active from 7 am to 10 pm daily, and another with full-day usage, covering 24 hours for bedrooms, kids' rooms, and studies. Distinct evaluation criteria for residential rooms that can be used as bedrooms are set, focusing specifically on **night-time** hours as is similarly suggested in CIBSE TM59. Unlike TM59, we suggest that not only **bedrooms but also kids' rooms and studies**, often used for sleeping purposes or with the potential to act as a bedroom, are included. The indoor operative temperature thresholds for sleeping areas are adapted to ensure acceptable sleep conditions, particularly during heat waves. These include both **integral** (Kelvin hours) **and absolute** (maximum temperature) **temperature limits during the night**. By extending the night criteria beyond bedrooms to other potential sleeping spaces, a comprehensive consideration of thermal comfort during critical night-time hours is aimed to be provided.

Instead of a two-stage indoor OA (proof fulfilled or not), **a three-stage assessment** with a **lower limit** indicating high heat resilience, an **upper limit** for acceptable heat resilience, and insufficient heat resilience above the upper limit is suggested. This structured approach provides a more holistic methodology for evaluating overheating risks and thermal comfort in residential buildings, encompassing general living areas as well as specific requirements for sleeping areas.

19.1   Explicit Indicator Set and Limits for Indoor Environments

A scheme of the developed indicator set shown in Table 19.1 provides a comprehensive overview of the criteria, thresholds, and processes involved. This indicator set comprises three criteria, of which only the first applies to the kitchen and living room:

- **Criteria I - Adaptive Integral Threshold (TWEH) During Occupancy** must be estimated for all rooms (living room, kitchen, bedroom, kid's room, and study) except bathrooms and floors (no overheating assessment for such room usages). TWEH is calculated above the adaptive temperature limit (according to EN 16798-1, Criteria I) for the entire year (integral threshold). The upper and lower acceptable limit values that must be complied with for Criteria I are depicted in Table 19.1.
- **Criteria II - Nighttime Criteria** apply only to bedrooms, kid's rooms, and studies, which can be transformed into sleeping spaces. Two specific criteria must be met for these rooms.
- **Criteria IIa – Adaptive Integral Threshold (TWEH) at Night** covers the TWEH above the adaptive temperature limit (according to EN 16798-1, Category I) and must be integrated from 10 PM to 7 AM throughout the year. The upper and lower limit values suggested in Table 19.1 must be fulfilled.
- **Criteria IIb - Absolute Exceedance Threshold of Indoor Temperature at Night** evaluates the maximum difference between the adaptive temperature limit (according to EN 16798-1, Category I) and the maximum recorded room temperature

between 10 PM and 7 AM. It must not exceed a specified temperature difference as defined in Table 19.1 throughout the year.

Tab. 19.1: Limits for the developed indicator set for indoor overheating assessment

	Description	Lower Limit (Kh/a)	Upper Limit (Kh/a)	Applicable Rooms	Time Frame
I	**Adaptive** Integral Threshold **for the whole Year**	**150** for Living Room & Kitchen, otherwise **240**	**300** for Living Room & Kitchen, otherwise **480**	Living Room, Kitchen, Bedroom, Kid's Room, Study	7 am to 10 pm resp. 24 h/d
IIa	**Adaptive** Integral Threshold at **Night**	45	90	Bedroom, Kid's Room, Study	10 pm to 7 am
IIb	**Absolute** Exceedance Threshold at **Night**	1 K	2 K	Bedroom, Kid's Room, Study	10 pm to 7 am

The assessment of the developed indicator set culminates in a three-tiered evaluation of overheating risk for each considered room, as detailed in the accompanying Figure

- A **Low overheating risk** indicates that all criteria are met at the defined lower limits.
- An **Acceptable overheating risk** assumes that all criteria are satisfied at least at the upper limits.
- A **High (unacceptable) overheating risk** certifies failure to meet at least one of the criteria.
- In the following, we discuss the suggested limits for the three criteria:
- The **lower and upper limits** for acceptable TWEH in **Criterion I** are derived from the assumption that a typical summer includes two heatwaves, each lasting five days. During these periods, a room temperature increase of 1 K (lower limit) or 2 K (upper limit) above the adaptive temperature limit is considered acceptable. This translates into a lower limit of 150 Kelvin-hours per annum (Kh/a) for kitchens and living rooms (calculated as 15 hours per day × 5 days × 1K × 2 heatwaves) and 240 Kh/a for bedrooms, children's rooms, and studies (24 hours per day × 5 days × 1K × 2 heatwaves). The upper limit is set to double these values since a 2 K exceedance during heatwaves is permitted.
- **Criterion IIa - Adaptive Integral Threshold at Night limits** are similar to Criterion I meaning the lower and upper limits of acceptable TWEH at night are based on an occupancy of only 9 hours (from 10 PM to 7 AM). During the two heatwaves, room temperatures are allowed an increase of 0.5 K (lower limit) or 1 K (upper limit) above the adaptive temperature limit for the entire night, resulting in limits of 45 Kh/a for the lower and 90 Kh/a for the upper limits.

– **_Criterion IIb - Absolute Exceedance Threshold at Night_** assumed lower and up-per **_limits_** of 1 K and 2 K based on building performance simulation tests. These tests have indicated that the selected Category I of the adaptive temperature limit in EN 16798-1 leads to high room temperatures during the night that might already be critical.

These limits are proposed as initial benchmarks and are subject to debate. They aim to integrate the benefits of various internationally available Overheating Assessment (OA) standards. Previous attempts to account for heat sensation influenced by humid air conditions have faced challenges, as standard Building Performance Simulation (BPS) tools do not consider moisture transport and storage within building components.

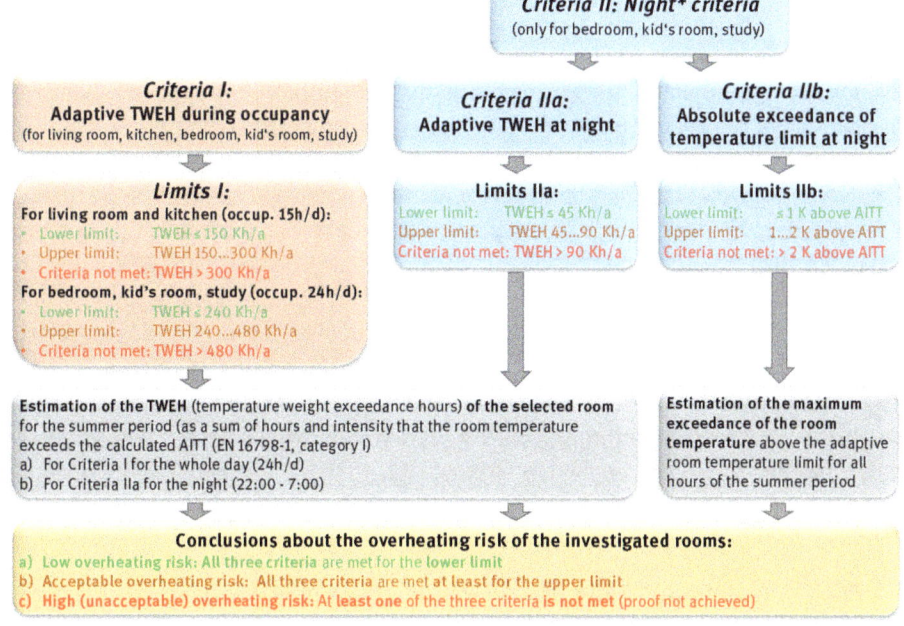

Fig. 19.1: Schematic of the developed indicator set for indoor overheating assessment (AITT is the adaptive indoor temperature threshold (EN 16798-1, category I), TWEH is the temperature weighted exceedance hours)

19.2   Application of the Refined Indicator Set

To test the proposed indicator set and compare the assessment results to existing standards we perform several BPS of a residential apartment building. The selected 'Gründerzeit' building was erected in the Wilhelminian period and is a frequent representative residential building type in Central European cities. For BPS the components of the buildings were represented as a three-dimensional model including all rooms. The buildings performance simulation (BPS) software IDA ICE 4.8 (EQUA, 2018) was used for the investigations considering wind- and temperature-gradient driven air exchange by window opening in a detailed manner (see Chapter Three of this book). An attic bedroom (eastern oriented) is chosen for BPS analysis which tends to show a considerable overheating risk because of the poor design of the subsequently converted attic. For the test of the indicator set, the building was virtually positioned in three locations in Germany with different summer conditions: Hamburg region (on the northern coast of Germany) as representative for cooler summers concerning the German average, Potsdam region (in the East of Germany) as a representative for average summers concerning Germany and Stuttgart region (in the South of Germany) as a representative for hot summers with respect the German average.

For these locations, measured weather data from the German Meteorological Service meteorological stations were used for an average present summer of the region. The procedure of identifying a year representing an average summer from the meteorological data from 1991-2020 is explained in detail by Kriesten et al. (2024). It is important to note that the meteorological stations are located outside the city and thus do not include urban climate effects. The statistical data of the identified average present summer of the three locations in Tab. 19.2 show the expected increase in warm nights (important for overheating risk evaluation because of different efficiency of nocturnal window ventilation), global irradiation, and maximum outdoor temperature from Hamburg over Potsdam to Stuttgart due to their differences in regional climate.

Tab. 19.2: Statistical data of the selected weather years for the summer period June 1st to August 31st

Location	Year	Average outdoor air temperature [°C]	Maximum outdoor air temperature [°C]	Number of hot days per year [-]	Number of tropical nights per year [-]	Number of warm nights per year [-]	Sum of global irradiation [kWh/m²]	Sum of direct solar irradiation [kWh/m²]	Sum of diffuse solar irradiation [kWh/m²]
Hamburg	2001	17.1	32.4	3	1	6	453	222	243
Potsdam	1997	19.0	33.2	10	0	7	490	263	227
Stuttgart	1991	19.0	35.6	8	4	11	504	273	232

The ***time-dependent adaptive indoor temperature threshold*** for all three loca-
tions/weather data was estimated according to EN 16798-1 (2019) using category I,
based on the outdoor air temperature course as hourly values. In Figure 19.2 the
course of the adaptive room temperature threshold and as a basis the outdoor air temperature is plotted for all three locations. The constant room temperature limit from
the German standard DIN 4108-2 is implemented for comparison. From the diagrams,
different observations can be made:

- The difference in indoor temperature threshold using an adaptive model compared to a constant value is remarkable, e.g. for Potsdam, the adaptive temperature limit ranges from 24.2 °C to 28.6 °C (in comparison to 26 °C as a constant
  value)
- Because the adaptive indoor temperature threshold is estimated by the daily outdoor temperature of the last seven days, a clear time delay from high outdoor
  temperature (heat waves) to high indoor temperature threshold can be observed.
  This should represent the inertia of the human adaptation process to changing
  environmental temperatures.
- The use of an adaptive indoor temperature threshold model also implies that the
  acceptable indoor temperature values at the beginning of summer (June) are
  mainly in the range of 24 °C to 26 °C whereas for the end of the summer (August)
  they are much higher between 26 °C and 29 °C. This considers the effect of human
  adaptation to environmental conditions in a more long-term manner and is
  mainly originated by higher nocturnal outdoor temperatures in August compared
  to June.
- Comparing the three locations, the maximum obtained adaptive indoor temperature limit of the summer is 28.1 °C for Hamburg, 28.6 °C for Potsdam, and 28.9 °C
  for Stuttgart which corresponds to the different summer conditions discussed
  above. However, such values are quite high although we already chose the strict-
  est category I of the adaptive model and not category II which is taken in CIBSE
  TM52 or TM59 meaning a 1 K higher indoor temperature threshold.

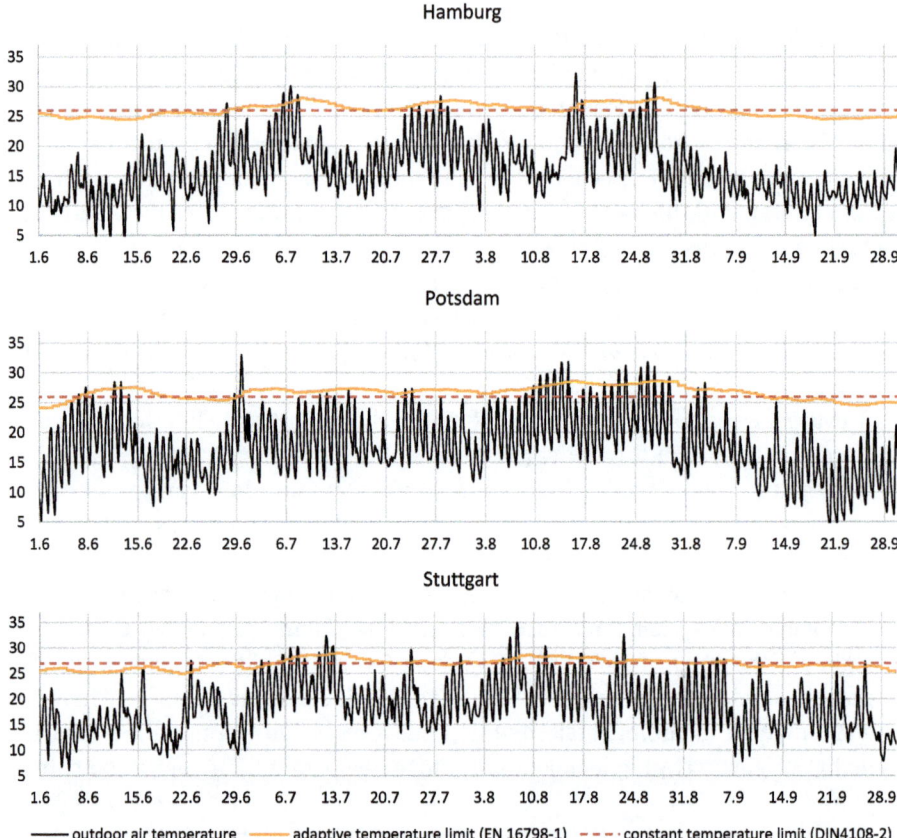

Fig. 19.2: Outdoor air temperature, adaptive indoor temperature threshold according to EN 16798-1, and constant indoor temperature threshold according to DIN 4108-2 (26 °C for Potsdam and Hamburg (summer climate region B), 27 °C for Stuttgart (summer climate region C)) for indoor operative temperatures for the selected locations (from June 1st to September 30th). All temperatures in °C.

The histograms of the calculated adaptive indoor temperature threshold values for the different locations in Figure 19.2 indicate a clear shift from lower acceptable indoor temperatures for the cooler summers in Hamburg to higher ones for hot summers in Stuttgart. Another observation is that the average of the adaptive temperature threshold (from June 1st to September 30th) is in the range of the constant temperature threshold defined in DIN 4108-2 (average of the adaptive threshold to the constant threshold: for Hamburg 26.1 °C to 26.0 °C; for Potsdam 26.7 °C to 26.0 °C; for Stuttgart 26.9 °C to 27.0 °C). This would be different when using category II in EN 16798-1 instead of category I implying a 1 K higher average of the adaptive limit which is another reason why we decided on category I for the adaptive temperature threshold estimation.

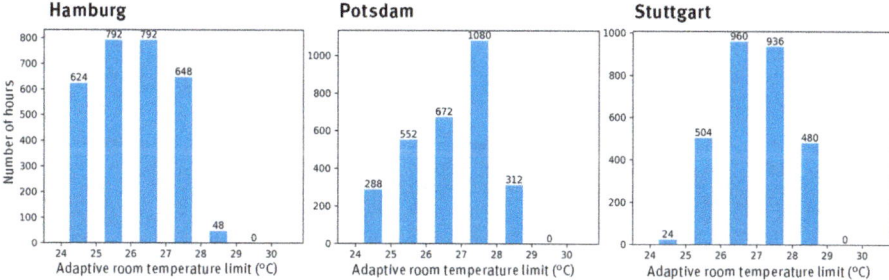

Fig. 19.3: Histograms of the adaptive indoor temperature threshold according to EN 16798-1 for the selected locations for the whole summer period (from June 1st to September 30th).

To test the proposed indicator set for its purpose on OA and compare the results with other OA approaches (like CIBSE TM52 or DIN 4108-2) we simulated the indoor temperature conditions of the attic bedroom in the 'Gründerzeit' residential building by BPS. Three simulation scenarios will be discussed which originate from the different weather data from the three different locations. The resulting operative indoor temperatures of the attic bedroom for these three scenarios are presented in Figure 19.3 along with the adaptive indoor temperature threshold (according to EN 16798-1) and the constant indoor temperature threshold (according to DIN 4108-2). The highest room temperatures of 33.0 °C are obtained for the hottest summer region of Stuttgart. In comparison, the maximum room temperature in the Hamburg region is only 29.8 °C. Because of the thermal inertia of the building and the fact that the adaptive temperature limit is calculated by the outdoor air temperature of the last seven days, the occurrence of the highest room temperatures coincides well with the highest acceptable (adaptive) indoor temperature thresholds. A comparison of the adaptive to the constant indoor temperature threshold given by DIN 4108-2 points out that the TWEH for the latter one is much higher because of the larger exceedance duration and height of the room temperature above the threshold.

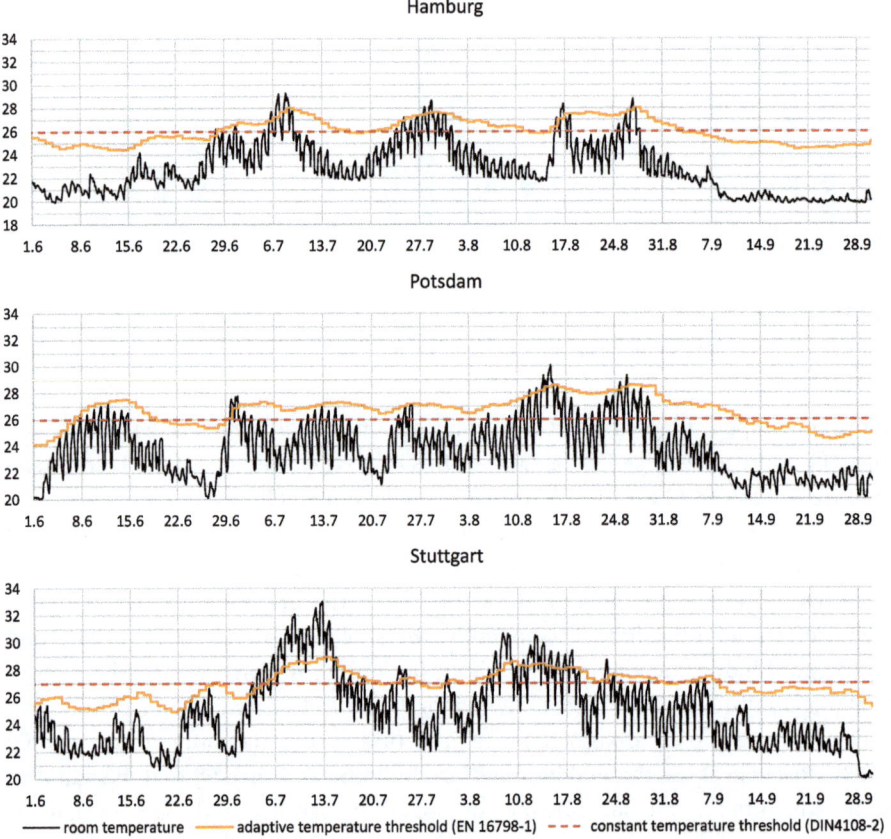

Fig. 19.4: Room temperature development of the apartment building (bedroom in the attic) for the three selected locations / weather data in Germany in correlation with constant indoor temperature threshold according to DIN 4108-2 and adaptive thresh-olds according to EN 16798-1.

For OA of the considered attic room the proposed *indicator set* as well as existing standards using constant temperature thresholds, DIN 4108-2 or CIBSE Guide A (2006) as well as adaptive temperature thresholds, CIBSE TM59, are *compared*. The results are summarised in Figure 19.4. For the regions of Hamburg and Potsdam the proposed indicator set results in a low overheating risk considering criteria I and IIa and acceptable overheating risk considering criteria IIb for the bedroom. The thresholds of the indicators from existing OA standards show a low overheating risk as well except for the CIBSE TM59 criterion 2. This night criterion demands that the operative temperature in the bedroom from 10 pm to 7 am shall not exceed 26 °C for more than 1% of annual hours. Also, the Criteria IIb of our proposed indicator states that the temperatures at night are high although not critical. The obtained maximum room temperature at night is found to be 28.9 °C for Hamburg and 29.0 °C for Potsdam which are not comfortable conditions for restful sleep. However, the CIBSE TM59 criterion 2 seems to be very strict only allowing 1 % of exceedance duration over 26 °C. It might be questionable if such high requirements are realistic to achieve for most bedrooms of free-running buildings without technical cooling. Such a hard criterion may lead to more bedrooms being actively cooled in the future. On the other hand, the bedrooms are the most critical rooms in residential buildings to ensure restful sleep and reduce critical health issues.

Overheating assessment criterias	Criteria thresholds		Simulation results		
	Lower threshold	Upper threshold	Hamburg	Potsdam	Stuttgart
Assesment by the proposed new indicator set:					
Criteria I: Adaptive TWEH during occupancy	TWEH max. 240 Kh/a	TWEH max. 480 Kh/a	80 Kh/a	50 Kh/a	650 Kh/a
Criteria IIa:Adaptive TWEH at night	TWEH max. 45 Kh/a	TWEH max. 90 Kh/a	10 Kh/a	10 Kh/a	140 Kh/a
Criteria IIb: Absolute exceedance of temperature threshold at night	1 K	2 K	1.1 K	1.2 K	3.2 K
Comparison to overheating criteria with constant temperature limit:					
DIN 4108-2	TWEH max. 1200 Kh/a above 25 °C / 26 °C / 27°C		370 Kh/a	520 Kh/a	1260 Kh/a
CIBSE Guide A 2006	exceedance hours[1] above 28 °C max. 1%		0,7%	0,9%	4,8%
Comparison to overheating criteria with adaptive temperature limit:					
CIBSE TM59 criterion 1	exceedance hours above adaptive limit[2] max. 3%		0,0%	0,0%	3,1%
CIBSE TM59 criterion 2 (night criterion)	exceedance hours[3] above 26 °C at night max 1%		2,6%	3,1%	8,6%

[1] number of hours where room temperature > 28 °C during occupancy (bedroom 24h/d) for the whole year

[2] number of hours where room temperature > 1 K above adaptive temperature limit during occupancy from May 1th to Sept. 30th

[3] number of hours where room temperature > 26 °C during night (10 pm to 7 am) for the whole year

Fig. 19.5: Room temperature development of the apartment building (bedroom in the attic) for the three selected locations / weather data in Germany in correlation with constant indoor temperature threshold according to DIN 4108-2 and adaptive thresh-olds according to EN 16798-1.

Considering the scenario that the building is located in Stuttgart, all OA criteria reveal that the bedroom has a high overheating risk and thus that heat adaptation measures are required. But while most criteria indicate a strong exceedance above the threshold, DIN 4108-2 and CIBSE TM59 criterion 1 are nearly fulfilled. This is in strong contrast to the significant exceedance of the threshold of CIBSE TM59 criterion 2 (night criteria) and CIBSE Guide A (2006). This difference also demonstrates that a suggested *multi-criteria approach enables a more holistic OA to check if a room is not only critical in general but for its specific purpose, e.g. as bedroom at night.*

In sum, the results of the proposed indicator set and thresholds showed its opportunities in this first test. However, further tests are necessary together with a more detailed discussion of what critical thresholds are in the context of overheating risk and health. The proposed indicator set serves as an initial impact to further develop OA by combining the strength from the different international existing indicators.

19.3   Limitations of the Refined Indicator Set

The analysis presented in the previous paragraphs illustrates that employing different OA standards results in divergent conclusions about the heat resilience of the assessed room. This variability can lead to confusion regarding the room's heat resilience, the adequacy of the building or room design, and the suitability of the indicator set for OA. The discrepancies arise primarily because different indicators focus on various aspects of overheating risk and occupancy types. Additionally, the use of diverse thresholds and limit values for OA significantly influences these outcomes. The process of selecting the most appropriate types of thresholds—whether integral, absolute, or combinations thereof—and developing verifiable limit values for OA is exceedingly complex. This complexity stems from numerous interrelated factors such as climate, human adaptability to temperature changes, gender, age, activity levels, clothing, and other conditions. As summarized in CIBSE TM52, all comfort standards encounter difficulties because they attempt to provide precise definitions for an inherently imprecise phenomenon. The adaptive comfort model in standard EN 16798-1, derived from extensive surveys, exemplifies this challenge as it does not distinguish between these influencing factors and often overlooks global differences. The debate over acceptable limits is driven by concerns about when indoor heat stress becomes detrimental to mental and physical health.

When addressing indoor overheating risks, it is crucial to consider the intended use of the building and the specific needs of its occupants. For residential buildings, this involves determining which rooms serve as bedrooms and their occupancy patterns. Differentiating between day and night use, as done in CIBSE TM59 and the proposed indicators, is beneficial for assessing the risk of overheating, which tends to be more critical at night for the recovery of mental and physical health during heat waves. The existing OA standards do not sufficiently differentiate between the needs of particularly vulnerable populations, such as elderly residents in nursing homes, who are more susceptible to heat stress. This gap highlights the need for further research to develop valid indicators that can reflect these varied needs effectively. The integration of humidity and sultriness in OA is also crucial, especially in locations with humid summer climates.

20 Conclusion

This chapter delves into various ***methodologies for assessing indoor overheating risk*** considering different kinds of thresholds of indicators, comparing them, and highlighting their strengths and limitations. It elucidates the ***fundamental systems behind the threshold definition***, such as absolute limits, model-based limits from adaptive approaches, or static approaches. Each of these methodologies has its own set of advantages and limitations, depending on the specific application context. Another essential aspect discussed in this chapter is the ***formulation of criteria for exceeding these thresholds***. The thresholds can be stringently defined to disallow any exceedances or can be designed to accommodate relative or integral exceedances. Such flexible thresholds allow for a more realistic assessment of thermal environmental conditions, accounting for fluctuations across daily or annual cycles. Determining the ***specific times when these thresholds are reviewed*** is also a critical factor. OA can be conducted throughout the year, on the most critical summer days, or at night, depending on the scenarios deemed most relevant. This is particularly important to meet the diverse requirements of users who may vary their activities and usage times of the building.

Based on these multifaceted considerations, a ***proposal for a refined approach for OA*** was developed. This refined indicator set integrates the benefits of existing approaches and expands them to include ***adaptive and user-specific criteria*** to facilitate a more comprehensive assessment of overheating risks. The proposed approach encompasses three criteria: Criterion I - Adaptive Integral Threshold (TWEH) during occupancy, Criterion IIa - Adaptive Integral Threshold at night, and Criterion IIb - Absolute Exceedance Threshold of indoor temperature at night. This approach was applied to a case study building and proved to be significantly more helpful than the known approaches.

In summary, while there is substantial potential in utilizing operative temperature and other refined indicators to enhance building simulation and overheating assessments, the complexity of human thermal perception—shaped by interrelated factors such as climate, culture, gender, age, activity, and clothing—necessitates ongoing research and development. This endeavor will ensure that future standards and building models effectively ***capture the nuanced realities of thermal comfort and health implications in diverse living and working environments***. This chapter paves the way for advancing methodologies by integrating strengths from various international standards and adapting them to the nuanced needs of different climatic environments.

21 References

1. ASHRAE (American Society of Heating, Refrigerating and Air-Conditioning Engineers). ASHRAE 55:2023 Thermal Environmental Conditions for Human Occupancy [Internet]. ASHRAE 55:2023 2023. Available from: https://www.ashrae.org/technical-resources/bookstore/standard-55-thermal-environmental-conditions-for-human-occupancy
2. Technical Committee ISO/TC 159/SC 5, International Organization for Standardization (ISO), International Organization for Standardization (ISO). ISO 7730:2005 Ergonomics of the thermal environment — Analytical determination and interpretation of thermal comfort using calculation of the PMV and PPD indices and local thermal comfort criteria [Internet]. ISO 7730:2005 Nov, 2005 p. 52. Available from: https://www.iso.org/standard/39155.html
3. Technical Committee ISO/TC 159/SC 5, International Organization for Standardization (ISO). ISO 7726:1998 Ergonomics of the thermal environment — Instruments for measuring physical quantities [Internet]. ISO 7726:1998 Nov, 1998 p. 51. Available from: https://www.iso.org/standard/14562.html
4. Kunkel S, Kontonasiou E, Arcipowska A, Mariottini F, Atanasiu B. Indoor Air Quality, Thermal Comfort and Daylight - Analysis of Residential Building Regulations in Eight EU member States. Buildings Performance Institute Europe (BPIE); 2015 Mar p. 1–100.
5. Luo M, Zhang H, Raftery P, Zhou L, Parkinson T, Arens E, et al. Detailed measured air speed distribution in four commercial buildings with ceiling fans. Build Environ. 2021 Aug 1;200:107979.
6. Duraković B, Mešetović S. Thermal performances of glazed energy storage systems with various storage materials: An experimental study. Sustain Cities Soc. 2019 Feb 1;45:422–30.
7. DeGreef JM, Chapman KS. Calculation of the Mean Radiant Temperature Directly Using Radiant Intensities. NIST [Internet]. 2017 Feb 19 [cited 2024 Feb 7]; Available from: https://www.nist.gov/publications/calculation-mean-radiant-temperature-directly-using-radiant-intensities
8. The Chartered Institution of Building Services Engineers (CIBSE). TM52: The Limits of Thermal Comfort: Avoiding Overheating in European Buildings [Internet]. TM52:2013 Oct, 2013 p. 24. Available from: https://www.cibse.org/knowledge-research/knowledge-portal/tm52-the-limits-of-thermal-comfort-avoiding-overheating-in-european-buildings
9. The Chartered Institution of Building Services Engineers (CIBSE). TM59 Design methodology for the assessment of overheating risk in homes [Internet]. TM59:2017 May, 2017 p. 17. Available from: https://www.cibse.org/knowledge-research/knowledge-portal/technical-memorandum-59-design-methodology-for-the-assessment-of-overheating-risk-in-homes
10. Deutsches Institut für Normung (DIN),. DIN 4108-2:2013-02, Wärmeschutz und Energie-Einsparung in Gebäuden_- Teil_2: Mindestanforderungen an den Wärmeschutz [Internet]. DIN4108-2:2013. Available from: https://www.beuth.de/de/-/-/167922321
11. SIA (Schweizerischer Ingenieur- und Architektenverein). SIA 180 / 2014 D - Wärmeschutz, Feuchteschutz und Raumklima in Gebäuden [Internet]. SIA180:2014 Jun, 2017 p. 72. Available from: http://shop.sia.ch
12. Austrian Standards International. ÖNORM B 8110-3 Wärmeschutz im Hochbau - Teil 3: Ermittlung der operativen Temperatur im Sommerfall (Parameter zur Vermeidung sommerlicher Überwärmung) [Internet]. ÖNORM B 8110-3:2020 Jun 1, 2020 p. 42.
13. Schweiker M, Kleber M, Wagner A. Long-term monitoring data from a naturally ventilated office building [Internet]. OSF; 2019 [cited 2024 Feb 11]. Available from: https://osf.io/2ydzg/
14. Schünemann C, Schiela D, Ortlepp R. Guidelines to Calibrate a Multi-Residential Building Simulation Model Addressing Overheating Evaluation and Residents' Influence. Buildings. 2021 Jun;11(6):242.

List of contributors

Peggy Freudenberg represents the Chair of Building Service and Climate-Responsive Architecture at Technische Universität Dresden (TUD). Since completing her Ph.D. in 2015, which focused on over-heating assessment of buildings, she has become a specialist in this field. In 2023, Freudenberg established and now leads the specialist group on overheating risk prediction within the WTA (Wissenschaftlich-Technische Arbeitsgemeinschaft für Bauwerkserhaltung und Denkmalpflege), emphasizing strategies to assess and mitigate overheating risk in existing buildings, a critical component in building conservation and heritage preservation. P. Freudenberg's career includes significant experience in third-party funded projects at TU Dresden, where she has applied her expertise in building physics to practical and theoretical challenges in the construction sector.

Christoph Schünemann is a Research Associate at the Leibniz Institute of Ecological Urban and Regional Development. He holds a PhD in Physics and has a diverse background in modeling and simulation ranging from System Dynamics modeling of complex social systems to Building Performance Simulation. The modeling work focuses on the efficacy evaluation of climate mitigation and adaptation measures – from a physical and a social perspective. In the HeatResilientCity project and further urban heat-related projects, Schünemann applies his expertise to adapt urban building and settlement structures, enhancing their climate resilience in response to rising urban temperatures. In other projects, including the PoliMod project, he analyses how social behavior hinders the implementation of climate mitigation and adaptation measures and how policies might affect such innovation diffusion processes.

Sabine Hoffmann is full professor at Rheinland-Pfälzischen Technischen Universität Kaiserslautern-Landau , specialized in Building Systems and Building Technology since 2014. Her expertise includes the integration of complex modelling systems into architectural design, particularly for energy-efficient and sustainable building practices. Hoffmann led the establishment of the Living Lab smart office space at the German Research Centre for Artificial Intelligence (DFKI) and played a crucial role in launching the master's program in Real Estate and Facilities - Management and Technology (IFMT). Her previous experience at the Lawrence Berkeley National Laboratory and the University of California's Centre for the Built Environment (CBE) further honed her skills in dynamic, switchable facade systems and thermal comfort modelling. Hoffmann's significant contributions to international projects and her role in developing software tools like SoloCalc underscore her capability in applying theoretical models to practical, real-world scenarios.

Abolfazl Ganji is project engineer at Transsolar Energietechnik GmbH in Stuttgart, a firm known for its innovative approaches to sustainable building technologies and climate-responsive architecture. With a strong educational background in Civil Engineering, Architecture, and Building Physics, Ganji effectively integrates these disciplines to develop new solutions aimed at enhancing health and comfort within buildings. His work during his postdoc at TU Kaiserslautern involved advanced simulation and optimization techniques, particularly in dynamic façade and complex fenestration systems.

Tim-Felix Kriesten is a PhD student at the Leibnitz Institute of Ecological Urban and Regional Development. He studied civil engineering at the TU Dresden.

Index